Revision & Practice

AQA GCSE 9–1 PHYSICS HIGHER

Series Editor: Primrose Kitten

Helen Reynolds
Alom Shaha

Contents

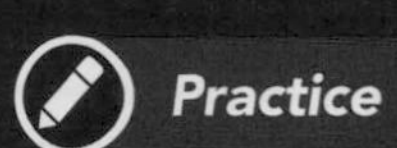

Shade in each level of the circle as you feel more confident and ready for your exam.

How to use this book

This book uses a three-step approach to revision: **Knowledge**, **Retrieval**, and **Practice**. It is important that you do all three; they work together to make your revision effective.

1 Knowledge

Knowledge comes first. Each chapter starts with a **Knowledge Organiser**. These are clear, easy-to-understand, concise summaries of the content that you need to know for your exam. The information is organised to show how one idea flows into the next so you can learn how all the science is tied together, rather than lots of disconnected facts.

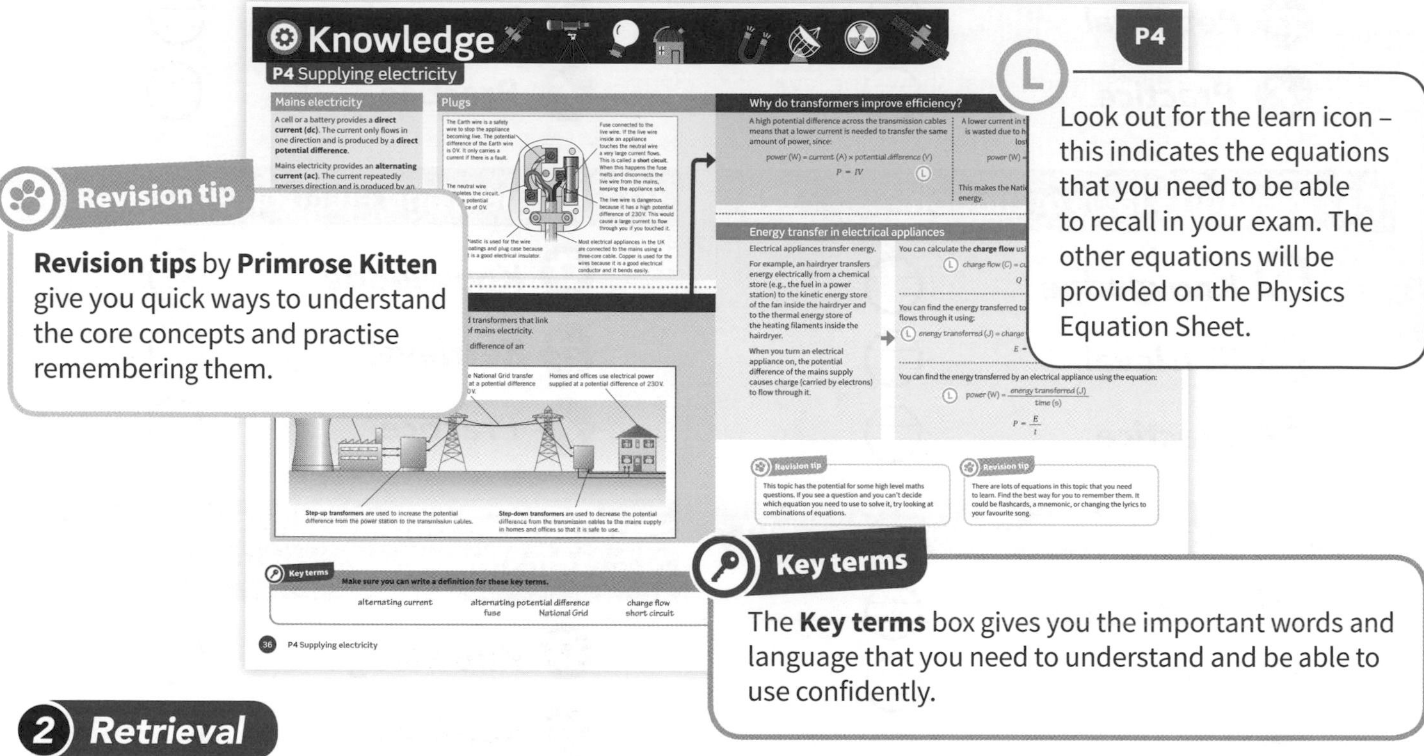

Knowledge
P4 Supplying electricity
P4
Mains electricity
Plugs
Why do transformers improve efficiency?
Energy transfer in electrical appliances
Revision tip
Revision tips by Primrose Kitten give you quick ways to understand the core concepts and practise remembering them.
Look out for the learn icon – this indicates the equations that you need to be able to recall in your exam. The other equations will be provided on the Physics Equation Sheet.
Key terms
Make sure you can write a definition for these key terms.
alternating current
alternating potential difference
fuse
National Grid
charge flow
short circuit
The Key terms box gives you the important words and language that you need to understand and be able to use confidently.
36 P4 Supplying electricity

2 Retrieval

The **Retrieval questions** help you learn and quickly recall the information you've acquired. These are short questions and answers about the content in the Knowledge Organiser. Cover up the answers with some paper; write down as many answers as you can from memory. Check back to the Knowledge Organiser for any you got wrong, then cover the answers and attempt *all* the questions again until you can answer all the questions correctly.

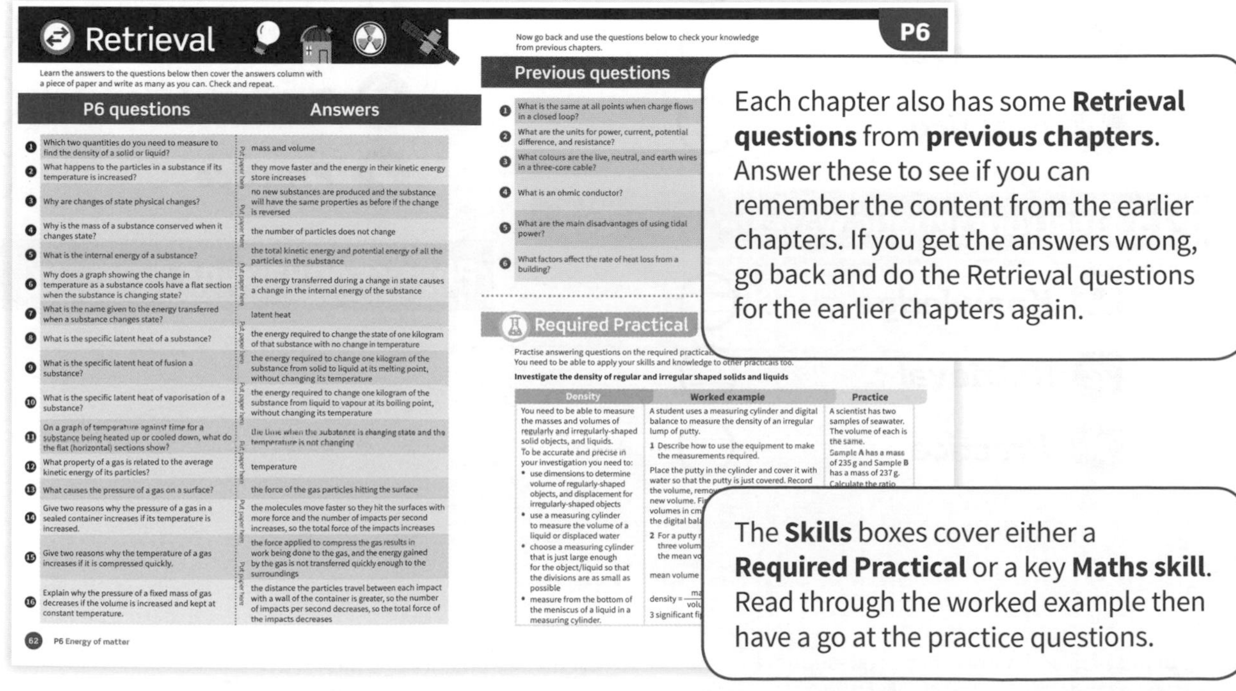

Retrieval
P6
Learn the answers to the questions below then cover the answers column with a piece of paper and write as many as you can. Check and repeat.
P6 questions
Answers
Now go back and use the questions below to check your knowledge from previous chapters.
Previous questions
Each chapter also has some Retrieval questions from previous chapters. Answer these to see if you can remember the content from the earlier chapters. If you get the answers wrong, go back and do the Retrieval questions for the earlier chapters again.
Required Practical
Investigate the density of regular and irregular shaped solids and liquids
Density
Worked example
Practice
The Skills boxes cover either a Required Practical or a key Maths skill. Read through the worked example then have a go at the practice questions.
62 P6 Energy of matter

Make sure you revisit the retrieval questions on different days to help them stick in your memory. You need to write down the answers each time, or say them out loud, otherwise it won't work.

3 Practice

Once you think you know the Knowledge Organiser and Retrieval answers really well you can move on to the final stage: **Practice**.

Each chapter has lots of **exam-style questions**, including some questions from previous chapters, to help you apply all the knowledge you have learnt and can retrieve.

Each question has a difficulty icon that shows the level of challenge.

 These questions build your confidence.

 These questions consolidate your knowledge.

 These questions stretch your understanding.

Make sure you attempt all of the questions no matter what grade you are aiming for.

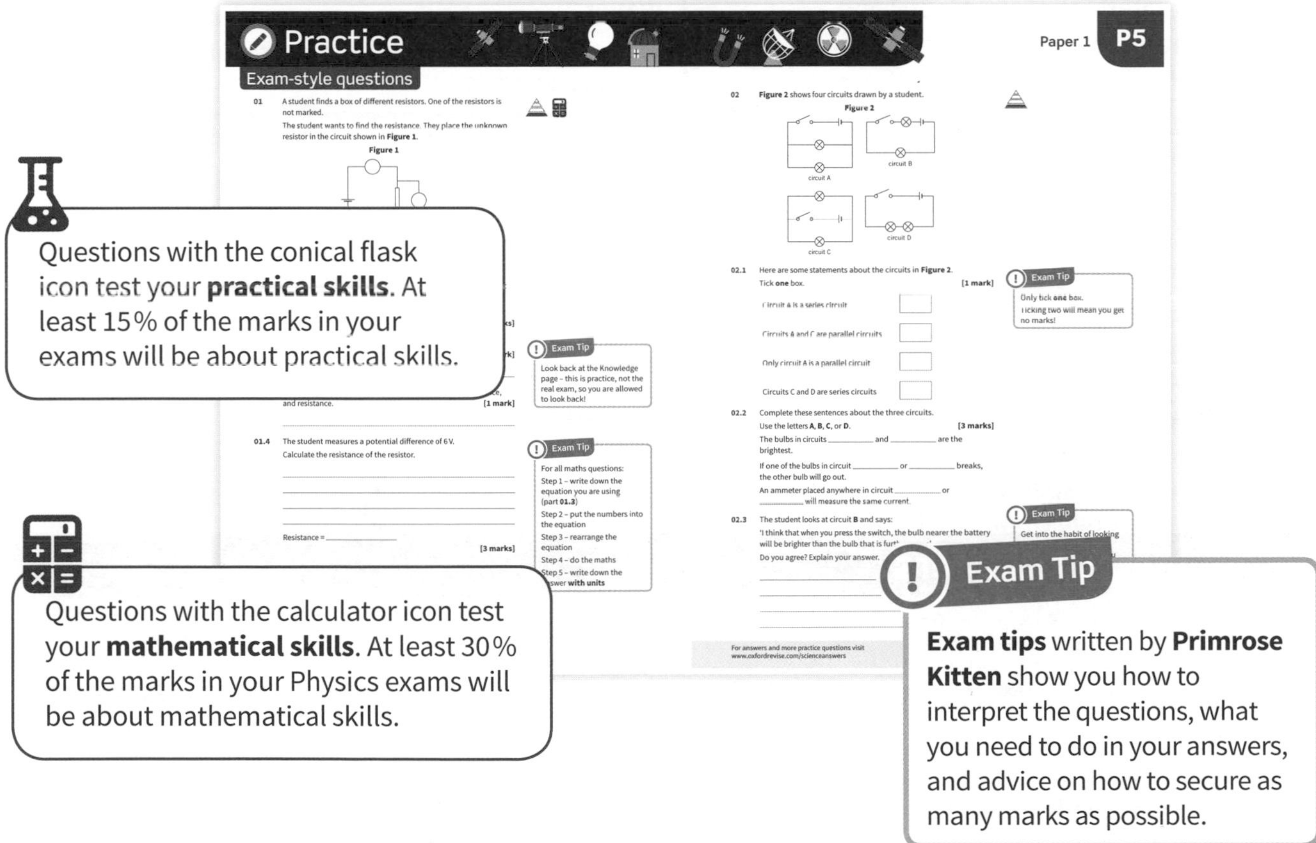
Practice
Exam-style questions
Paper 1
P5
Questions with the conical flask icon test your **practical skills**. At least 15% of the marks in your exams will be about practical skills.
Questions with the calculator icon test your **mathematical skills**. At least 30% of the marks in your Physics exams will be about mathematical skills.
Exam Tip
Exam tips written by **Primrose Kitten** show you how to interpret the questions, what you need to do in your answers, and advice on how to secure as many marks as possible.

kerboodle

All the **answers** are on Kerboodle and the website, along with even more exam-style questions. www.oxfordrevise.com/scienceanswers

Knowledge

P1 Energy stores and transfers

Systems

A **system** is an object or group of objects.

Whenever anything changes in a system, energy is transferred between its stores or to the surroundings.

A **closed system** is one where no energy can escape to or enter from the surroundings. The total energy in a closed system never changes.

Energy stores

kinetic	energy an object has because it is moving
gravitational potential	energy an object has because of its height above the ground
elastic potential	energy an elastic object has when it is stretched or compressed
thermal (or internal)	energy an object has because of its temperature (the total kinetic and potential energy of the particles in the object)
chemical	energy that can be transferred by chemical reactions involving foods, fuels, and the chemicals in batteries
nuclear	energy stored in the nucleus of an atom
magnetic	energy a magnetic object has when it is near a magnet or in a magnetic field
electrostatic	energy a charged object has when near another charged object

Energy transfers

Energy can be transferred to and from different stores by:

Heating

Energy is transferred from one object to another object with a lower temperature.

Waves

Waves (e.g., light and sound waves) can transfer energy by radiation.

Electricity

When an electric current flows it can transfer energy.

Forces (mechanical work)

Energy is transferred when a force moves or changes the shape of an object.

Examples of energy transfers

When you stretch a rubber band, energy from your chemical store is mechanically transferred to the rubber band's elastic potential store.

When a block is dropped from a height, energy is mechanically transferred (by the force of gravity) from the block's gravitational potential store to its kinetic store.

When this block hits the ground, energy from its kinetic energy store is transferred mechanically and by sound waves to the thermal energy store of the surroundings.

The electric current in a kettle transfers energy to the heating element's thermal energy store. Energy is then transferred by heating from the heating element's thermal energy store to the thermal energy store of the water.

When an object slows down due to friction, energy is mechanically transferred from the object's kinetic store to its thermal store, the thermal store of the object it is rubbing against, and to the surroundings.

Energy equations

An object's gravitational potential energy store depends on its height above the ground, the gravitational field strength, and its mass.

gravitational potential energy (J) = mass (kg) × gravitational field strength (N/kg) × height (m)

$$E_p = m\,g\,h$$

An object's kinetic energy store depends only on its mass and speed.

kinetic energy (J) = 0.5 × mass (kg) × (speed)^2 (m/s)

$$E_k = \frac{1}{2} m\,v^2$$

Power is how much work is done (or how much energy is transferred) per second. Work done means the same thing as energy transferred. The unit of power is the watt (W).

1 watt = 1 joule of energy transferred per second

$$\text{power (W)} = \frac{\text{energy transferred (J)}}{\text{time (s)}}$$

$$P = \frac{E}{t}$$

or

$$\text{power (W)} = \frac{\text{work done (J)}}{\text{time (s)}}$$

$$P = \frac{W}{t}$$

The elastic potential energy store of a stretched spring can be calculated using:

elastic potential energy (J) = 0.5 × spring constant (N/m) × (extension)^2 (m)

$E_e = \frac{1}{2} k\,e^2$ (assuming the limit of proportionality has not been exceeded)

This equation will be given to you on the equation sheet, but you need to be able to select and apply it to the correct questions.

Useful and dissipated energy

Energy cannot be created or destroyed – it can only be transferred usefully, stored, or dissipated (wasted).

Dissipated energy is often described as being wasted.

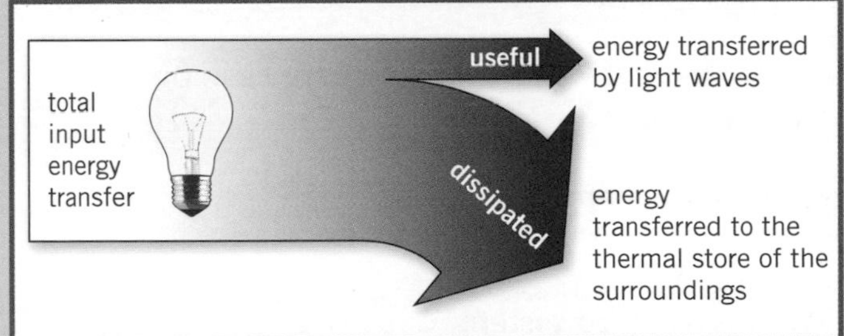

Energy is never entirely transferred usefully – some energy is always dissipated, meaning it is transferred to less useful stores.

All energy eventually ends up transferred to the thermal energy store of the surroundings.

In machines, work done against the force of friction usually causes energy to be wasted because energy is transferred to the thermal store of the machine and its surroundings.

Lubrication is a way of reducing unwanted energy transfer due to friction.

Streamlining is a way of reducing energy wasted due to air resistance or drag in water.

Use of thermal insulation is a way of reducing energy wasted due to heat dissipated to the surroundings.

Efficiency is a measure of how much energy is transferred usefully. You must know the equation to calculate efficiency as a *decimal*:

$$\text{efficiency} = \frac{\text{useful output energy transfer (J)}}{\text{total input energy transfer (J)}}$$

or

$$\text{efficiency} = \frac{\text{useful power output (W)}}{\text{total power input (W)}}$$

To give efficiency as a *percentage*, just multiply the result from the above calculation by 100 and add the % sign to the answer.

Key terms

Make sure you can write a definition for these key terms.

chemical · closed system · dissipated · efficiency · elastic potential · electrostatic · gravitational potential · kinetic · lubrication · magnetic · nuclear · power · streamlining · system · thermal · work done

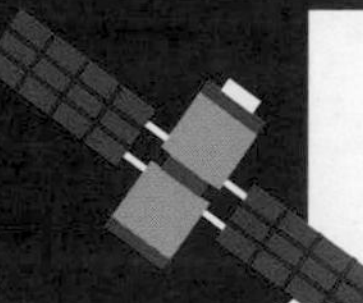

Retrieval

Learn the answers to the questions below then cover the answers column with a piece of paper and write as many as you can. Check and repeat.

	P1 questions	Answers
1	Name the eight energy stores.	kinetic, gravitational potential, elastic potential, thermal, chemical, nuclear, magnetic, electrostatic
2	Name the four ways in which energy can be transferred.	heating, waves, electric current, mechanically (by forces)
3	What is a system?	an object or group of objects
4	What is a closed system?	a system where no energy can be transferred to or from the surroundings – the total energy in the system stays the same
5	What is work done?	energy transferred when a force moves an object
6	What is the unit for energy?	joules (J)
7	What is one joule of work?	the work done when a force of 1 N causes an object to move 1 m in the direction of the force
8	Describe the energy transfer when a moving car slows down.	energy is transferred mechanically from the kinetic store of the car to the thermal store of its brakes. Some energy is dissipated to the thermal store of the surroundings
9	Describe the energy transfer when an electric kettle is used to heat water.	the electric current in a kettle transfers energy to the heating element's thermal store – energy is then transferred by heating from the heating element's thermal store to the thermal store of the water
10	Describe the energy transfer when a ball is fired using an elastic band.	energy is transferred mechanically from the elastic store of the elastic band to the kinetic store of the band – some energy is dissipated to the thermal store of the surroundings
11	Describe the energy transfer when a battery powered toy car is used.	energy is transferred electrically from the chemical store of the battery to the kinetic store of the toy car – some energy is dissipated to the thermal store of the surroundings
12	Describe the energy transfer when a falling apple hits the ground.	energy is transferred from the kinetic store of the apple and dissipated to the thermal store of the surroundings by sound waves
13	Name the unit that represents one joule transferred per second.	watt (W)

Put paper here

Maths Skills

Practise your maths skills using the worked example and practice questions below.

Rearranging equations	Worked example	Practice
You need to be able to rearrange and apply many equations in physics, for example, the equation for power. Power is the rate at which energy is transferred or the rate at which work is done. It can be calculated using: $\text{power (W)} = \frac{\text{energy transferred (J)}}{\text{time taken (s)}}$ Remember: • a power of 1 W means a rate of energy transfer of 1 J per second • the shorter the time taken for an energy transfer, the greater the power • if the time taken to transfer energy is given in minutes or hours, you must convert it to seconds before calculating the power.	A microwave is marked as having a power of 900 W. How long does it take to transfer 13 500 J of energy? **Step 1:** write down the equation. $\text{power (W)} = \frac{\text{energy transferred (J)}}{\text{time taken (s)}}$ **Step 2:** work out which information in the question relates to the variables in the equation. power = 900 W, energy transferred = 13 500 J **Step 3:** put the numbers into the equation. $900 = \frac{13\,500}{\text{time}}$ **Step 4:** rearrange the equation. Multiply both sides of the equation by t: $900 \times \text{time} = 13\,500$ Then divide both sides of the equation by 900: $\text{time} = \frac{13\,500}{900} = 15\text{ s}$ *Alternatively:* **Step 3:** rearrange the equation first. $\text{time} = \frac{\text{energy}}{\text{power}}$ **Step 4:** put the numbers into the equation. $= \frac{13\,500}{900} = 15\text{ s}$	**1** An LED lamp transfers 360 J of energy in 30 seconds. Calculate its power. **2** An electric cooker transfers 3 MJ of energy in 5 minutes. Calculate its power. **3** An electric kettle has a power of 3000 W. How much energy does it transfer in 45 seconds? **4** Rearrange the equation linking efficiency, useful power output, and total power input to calculate the total power output of a system.

Practice

Exam-style questions

01 A student is playing with a slinky spring.

She holds one end and pulls the other end of the spring.

01.1 There is energy in the elastic store of the spring.

Identify the energy store that had more energy in it before she pulled the spring, and where this energy has come from. **[2 marks]**

____

____

____

01.2 The spring has a spring constant of 20 N/m.

When the student pulls the spring it extends by 0.2 m.

Calculate the energy stored in the spring. Use the correct equation from the *Physics Equations Sheet*. **[2 marks]**

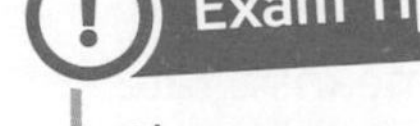

Always start your calculations by writing down the equation you're going to use!

____

____

Energy stored = ____________________ J

02 This question is about conservation of energy.

02.1 Complete the statement of the law of conservation of energy. **[1 mark]**

Energy cannot be ____

____________________ only transferred, stored, or dissipated.

02.2 When you consider energy changes, it is helpful to talk about the system that you are considering.

Define what a closed system is. **[1 mark]**

Key definitions are really important to learn as they are easy marks in the exam.

____

____

02.3 A pendulum swings backwards and forwards. The pendulum gradually stops swinging.

Circle the correct bold word or words in this sentence.

A pendulum **is / is not** a closed system. **[1 mark]**

02.4 Explain why you have selected the word or words in this sentence. **[2 marks]**

03 A student investigates the energy transfers of a tennis ball when it is dropped onto the classroom floor.

She wants to calculate the energy transferred to the floor and surroundings when the ball first bounces.

03.1 Describe the measurements that she needs to make.

Suggest measuring instruments that she could use to do this. **[5 marks]**

> **Exam Tip**
> There are two parts to this question. For everything you want to measure you also have to say how you are going to measure it.

03.2 Suggest a problem that she may have with making the measurements and how it might be overcome. **[2 marks]**

03.3 Describe in detail how she can use the measurements to calculate the energy transferred to the floor/surroundings. **[5 marks]**

> **Exam Tip**
> The majority of the time problems in practicals can be over come by using technology.

03.4 Suggest whether you can or cannot use the term efficiency when describing what happens when the ball bounces.

Justify your answer. **[2 marks]**

04 A gymnast runs and lands on a springboard. Springs store energy when they are stretched or compressed. You can use the same equation to calculate energy in a stretched or compressed spring. You use compression instead of extension in your equation.

04.1 Write down the equation that links kinetic energy, mass, and speed. **[1 mark]**

04.2 The gymnast is travelling at 10 m/s when she lands on the springboard. The springboard contains a large spring that compresses when she lands on it. The mass of the gymnast is 40 kg. Calculate the kinetic energy of the gymnast. **[2 marks]**

04.3 The spring constant of the spring is 20 000 N/m. Assume that all of the energy in the kinetic energy store is transferred to the elastic potential energy store of the spring when the gymnast lands. Calculate the compression of the spring. **[3 marks]**

04.4 A manufacturer tests the spring in a springboard. She finds that the spring does not compress by the distance predicted by the calculation. Suggest whether the actual compression is bigger or smaller than the predicted compression. Explain your answer. **[3 marks]**

> **Exam Tip**
>
> If you write down the symbol equation, ensure you write it down carefully and don't get the symbols mixed up with other. It's ok to write the word equation down if you'll find it less confusing.

> **Exam Tip**
>
> If you're confused, go back and read the hints given at the start of the question.

05 An Olympic archer can shoot an arrow at a target that is 80 m away. The archer draws back the string of a bow and releases it to shoot the arrow.

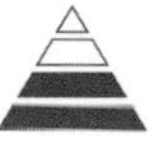

05.1 Describe the energy changes between the moment just before the archer releases the arrow and the moment when the arrow no longer has contact with the string. Write down the method by which the energy is transferred. **[3 marks]**

05.2 The archer decides to calculate the energy stored in the string. She makes the following estimates:

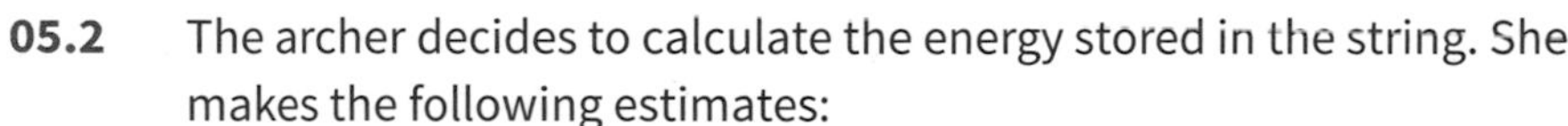

- the spring constant of the string is 10^5 N/kg
- the extension of the string is 5 cm.

Use the correct equation from the *Physics Equations Sheet* to calculate the energy stored in the string. **[2 marks]**

> **Exam Tip**
>
> Don't worry about the big numbers in this calculation. If you're not sure how to put 10^5 into your calculator then now is a great chance to practise.

05.3 The archer wants to find the speed of the arrow when it leaves the bow by direct measurement. Suggest a technique for finding the speed of the arrow. **[1 mark]**

05.4 Suggest **one** reason why the speed of the arrow determined by this technique might be lower than the values from **05.2** suggest. **[1 mark]**

05.5 The arrow hits the target and stops. Give the name of the energy store that has gained energy as a result of this process. **[1 mark]**

06 The high-speed train in **Figure 1** travels at a top speed of 200 mph (90 m/s)

Figure 1

06.1 Suggest in terms of energy why the front of the train has the shape shown in **Figure 1**. **[1 mark]**

06.2 Write down the equation that links kinetic energy, mass, and speed. **[1 mark]**

06.3 Calculate the kinetic energy of the train when it is travelling at top speed. Write your answer in kJ. Give your answer to three significant figures. The mass of the train is 700 000 kg. **[4 marks]**

> **Exam Tip**
> Be careful with this equation, it's just the speed that is squared not the whole right-hand side.

06.4 Write down the work done by the brakes to bring the train to a complete stop. **[1 mark]**

07 An astronaut on the Moon is holding a hammer. He lets it go and it hits the ground.

07.1 Describe the energy changes from the moment when he drops the hammer to the moment before the hammer hits the ground. **[2 marks]**

> **Exam Tip**
> The energy at the start and the end will be different, be sure to use the correct terms to describe them.

07.2 Give the method of energy transfer that is producing the change. **[1 mark]**

07.3 Compare the speed of the hammer when it hits the ground on the Moon with the speed that it would hit the ground if he was on Earth.

- Assume the mass of the hammer is the same.
- The gravitational field strength of the Moon is less than that of Earth.
- The hammer is dropped from the same height.

Explain your answer. You do not need to do any calculations. **[5 marks]**

> **Exam Tip**
> The question has given you three bullet points to discuss – make sure you cover all of them in your answer.

08 A tall hotel building has a lift designed to carry guests and their luggage.

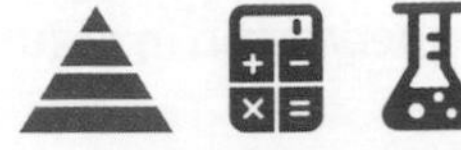

08.1 Write down the equation that links gravitational potential energy, mass, gravitational field strength, and height. **[1 mark]**

08.2 A family, the lift, and their luggage have a total mass of 1220 kg. Calculate the change in the gravitational potential energy store when the family moves up four floors. Each floor is 3.0 m high. Gravitational field strength is 9.8 N/kg. **[2 marks]**

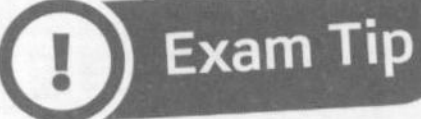

Exam Tip

Don't forget to calculate the total height first!

08.3 Write down the equation for efficiency. **[1 mark]**

08.4 The motor transfers a total energy of 280 kJ. Calculate the efficiency of the lift motor as a percentage. **[3 marks]**

Exam Tip

Students are always forgetting to convert the decimal to a percentage or forgetting the % sign after the answer, both of these are mistakes that will cost you marks.

08.5 There is also a lift designed to carry trolleys of cleaning supplies and laundry. The power of this lift is 10 kW. Suggest measurements that you could make to work out which lift motor is more efficient. Explain how you would use the measurements to calculate the efficiency. **[6 marks]**

08.6 Suggest why there would be uncertainty in your calculation of efficiency. **[1 mark]**

09 In a factory, a forklift truck does work lifting a box onto a shelf.

09.1 Write down the equation that links power, energy transferred, and time. **[1 mark]**

09.2 To lift the box onto the shelf, the motor of the forklift truck transfers 30 000 J of energy. The power of the truck motor is 15 000 W. Show that it takes 2 seconds for the truck to lift the box onto the shelf. **[3 marks]**

Exam Tip

'Show' questions are amazing – you already know what the answer is, you just need to do the working. The great thing is if you don't get the right answer you can just try again!

09.3 A second truck lifts an identical box onto the same shelf. The second truck takes longer to move the box onto the shelf. Compare the power of the two trucks. **[1 mark]**

09.4 Write down the name of the store that has more energy in it after the truck has moved the box. **[1 mark]**

10 A teacher set the students a challenge to check their understanding of energy. She puts a piece of track on the desk and raises one end. A marble rolls down the track and moves horizontally off the desk, as shown in **Figure 2**.

Figure 2

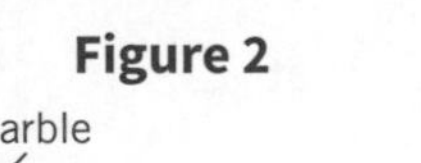

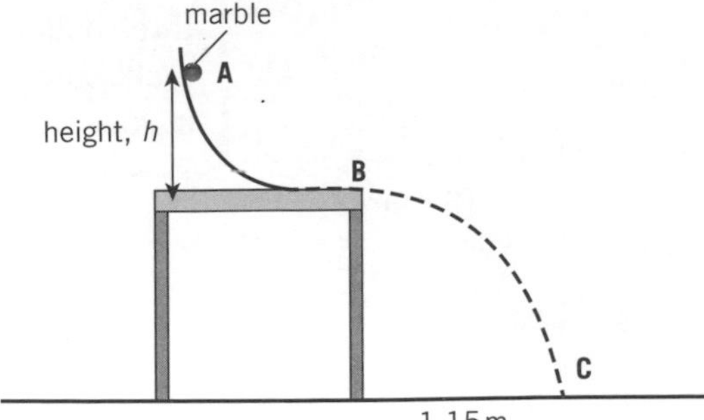

Exam Tip

Try labelling the diagram at each point with the types of energy, it will help with later questions.

10.1 Describe the changes in energy between points **A** and **B**. **[2 marks]**

10.2 The ball is travelling at a horizontal speed of 1.3 m/s when it leaves the table.

Suggest the equipment that the students could use to measure the speed of the ball at the end of the track. **[1 mark]**

10.3 There is a target at point **C** on the floor that the students are aiming to hit. The horizontal distance of C from the edge of the table is 1.15 m. It takes 820 ms for the marble to move from **B** to **C**. Calculate the distance travelled by the ball.

Write down **one** assumption that you made. Write down whether or not it will hit the target. **[5 marks]**

There are three different parts to this question, make sure you include all of them in your answer.

10.4 Suggest how the students should adjust the equipment to make the ball hit the target. Explain your answer in terms of energy. **[4 marks]**

11 A student researches the efficiency of different electric motors used in cars. In all system changes energy is wasted.

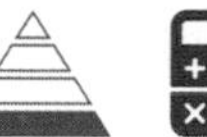

11.1 Describe what is meant by wasted energy. **[1 mark]**

11.2 The student knows that car **A** requires a total energy input of 20 J to move across the floor. It has useful energy output of 12 J. Write down the equation that links efficiency, total output energy, and total input energy. **[1 mark]**

11.3 Calculate the efficiency of car **A** as a decimal. **[2 marks]**

This question has asked for the answer as a decimal, so don't multiply by 100.

11.4 The efficiency of car **B** is 50%. Write down which car wastes more energy. Explain your answer. **[2 marks]**

12 A student investigates how the height of a ramp affects the speed of a trolley. He puts the trolley at the top of a ramp. He releases the trolley and measures the speed of the trolley at the bottom of the ramp.

12.1 Explain why the student cannot use a ruler and stopclock to measure the speed at the bottom of the ramp.

Suggest the equipment that the student should use instead. **[2 marks]**

12.2 The student measures the height of the ramp. It is 0.12 m. Gravitational field strength is 9.8 N/kg. The mass of the trolley is 0.25 kg. Write down the equation that links gravitational potential energy, mass, gravitational field strength, and height. **[1 mark]**

12.3 Calculate the gravitational potential energy of the trolley at the top of the ramp. **[2 marks]**

Exam Tip

You'll always be given the value for gravitational field strength, so you don't need to remember it.

12.4 The student measures the kinetic energy of the trolley at the bottom of the ramp. He discovers that the kinetic energy of the trolley is less than the gravitational potential energy.

He says: 'There is a mistake in the data. The answers should be the same.'

Do you agree? Give reasons for your answer. **[2 marks]**

13 A student is watching a video about car safety. In the video, cars with robot drivers hit a barrier. The student notices that:

- some cars stop after colliding with a barrier
- some cars bounce off the barrier.

13.1 The barrier compresses on impact but does not return to its original shape.

Explain why you cannot use the equation for elastic potential energy in energy transfer calculations in this case. **[1 mark]**

13.2 Data about two cars colliding with the same barrier is shown in **Table 1**.

Table 1

Car	Mass in kg	Speed before hitting the barrier in m/s	Speed after hitting the barrier in m/s	Energy transferred during collision in J
A	1000	50	0	1.25×10^6
B	1000	50	25	

Calculate the energy transferred by car **B** during the collision.

Give your answer in standard form and to an appropriate number of significant figures. **[7 marks]**

> **Exam Tip**
> Don't use more significant figures than the question asks for.

13.3 Identify which car transfers less energy to the surroundings. Explain why. **[2 marks]**

> **Exam Tip**
> You need to give the why to get the second mark in this question.

13.4 Another student looks at the data. He sees that the speed of car **B** has been reduced by 50% compared with car **A** after the collision. Car **A**'s speed has been reduced by 100% after the collision. He says that he expected the energy transferred by car **B** would be half the energy transferred by car **A**.

Suggest why this is **not** the case. **[2 marks]**

> **Exam Tip**
> There are lots of other energy types involved in this situation – try to think of as many as you can.

14 A motorcycle company produces data that shows how the efficiency of a motorcycle compares with that of a car over a range of speeds. Graphs of efficiency against speed for the motorcycle and car are shown in **Figure 3**.

Figure 3

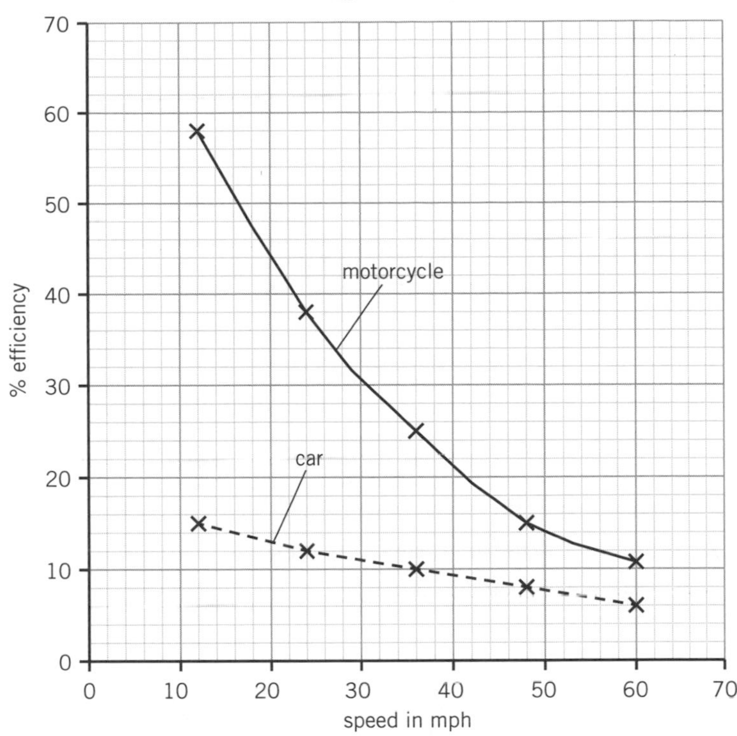

Efficiency is a measure of the transfer of input energy to useful energy.

14.1 Suggest what the company used as the source of input energy. **[1 mark]**

14.2 Write down the energy store into which this energy was usefully transferred. **[1 mark]**

14.3 Suggest a process that transfers energy to a store that is **not** useful. **[2 marks]**

14.4 Describe how the efficiency of each vehicle varies with speed. **[4 marks]**

14.5 Calculate the rate of change of efficiency with speed for the motorcycle at 30 mph.

The units are % per mph. **[3 marks]**

Exam Tip

The units give you a big clue to the equation. % per mph is a bit like saying % divided by mph.

P2 Energy transfers by heating

Thermal conductivity

The **thermal conductivity** of a material tells you how quickly energy is transmitted through it by thermal conduction.

You can test the thermal conductivity of rods made of different metals using this experimental set-up. Each rod must have the same diameter and length, and the same temperature difference between its ends.

One end of each rod is covered in wax and the other ends are heated equally. The faster the wax melts, the higher the thermal conductivity of the metal.

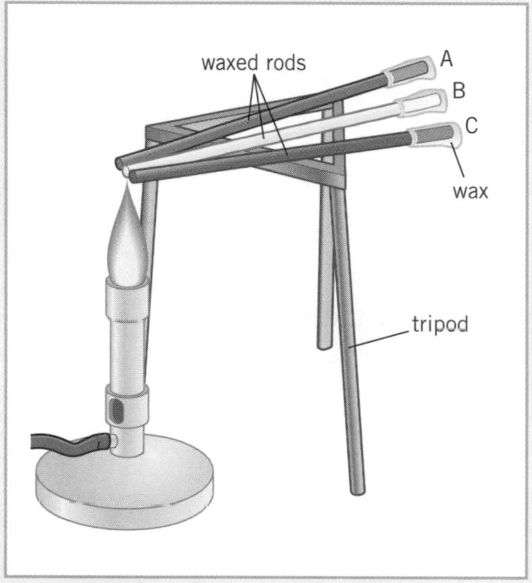

Insulating buildings

Heating bills can be expensive so it is important to reduce the rate of heat loss from buildings.

Some factors that affect the rate of heat loss from a building include:

1 the thickness of its walls and roof
2 the thermal conductivity of its walls and roof.
lower thermal conductivity = lower rate of heat loss

The thermal conductivity of the walls and roof can be reduced by using **thermal insulators**.

A thermal insulator is a material which has a low thermal conductivity. The rate of energy transfer through an insulator is low.

The energy transfer per second through a material depends on:

1 the material's thermal conductivity
2 the temperature difference between the two sides of the material
3 the thickness of the material.

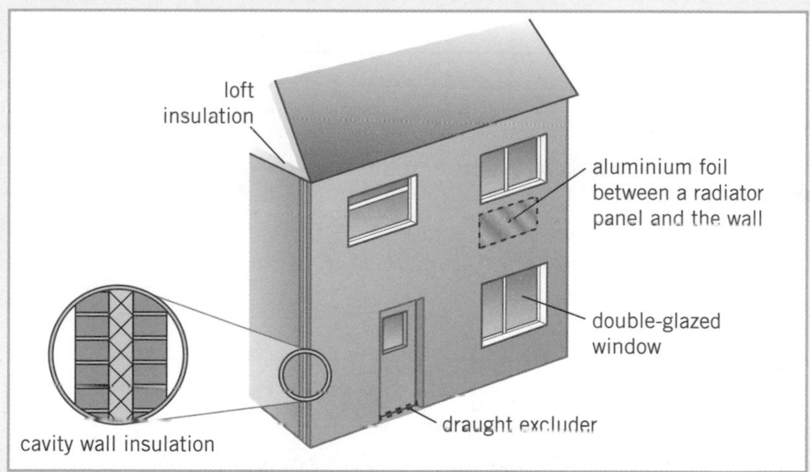

Specific heat capacity

When a substance is heated or cooled the temperature change depends on:

- the substance's mass
- the type of material
- how much energy is transferred to it.

Every type of material has a **specific heat capacity** – the amount of energy needed to raise the temperature of 1 kg of the substance by 1 °C.

The energy transferred to the thermal store of a substance can be calculated from the substance's mass, specific heat capacity, and temperature change:

change in thermal energy (J) = mass (kg) × specific heat capacity (J/kg°C) × temperature change (°C)

$$\Delta E = m\ c\ \Delta\theta$$

This equation will be given to you on the equation sheet, but you need to be able to select and apply it to the correct questions.

Key terms

Make sure you can write a definition for these key terms.

absorb   black body   electromagnetic spectrum   emit   greenhouse gas

Infrared radiation

Infrared radiation is part of the **electromagnetic spectrum**.

All objects **emit** (give out) and **absorb** (take in) infrared radiation.

The higher the temperature of an object, the more infrared radiation it emits in a given time.

A good absorber of infrared radiation is also a good emitter.

For an object at a constant temperature:

- infrared radiation emitted = infrared radiation absorbed
- infrared radiation is emitted across a continuous range of wavelengths.

An object's temperature will increase if it absorbs infrared radiation at a higher rate than it emits it. This rule applies to the planet Earth.

Black bodies

A **black body** is a theoretical object that absorbs 100% of the radiation that falls on it.

A perfect black body would not reflect or transmit any radiation, and would also be a perfect emitter of radiation.

Radiation and the Earth's temperature

The temperature of the Earth depends on lots of factors, including the rate at which visible light and infrared radiation are reflected, absorbed, and emitted by the Earth's atmosphere and surface.

Greenhouse gases absorb infrared radiation emitted by the surface of the Earth and prevent it escaping into space. They then re-emit the infrared radiation back towards the surface of the Earth, increasing the Earth's temperature. Greenhouse gases in the Earth's atmosphere include water vapour, methane, and carbon dioxide.

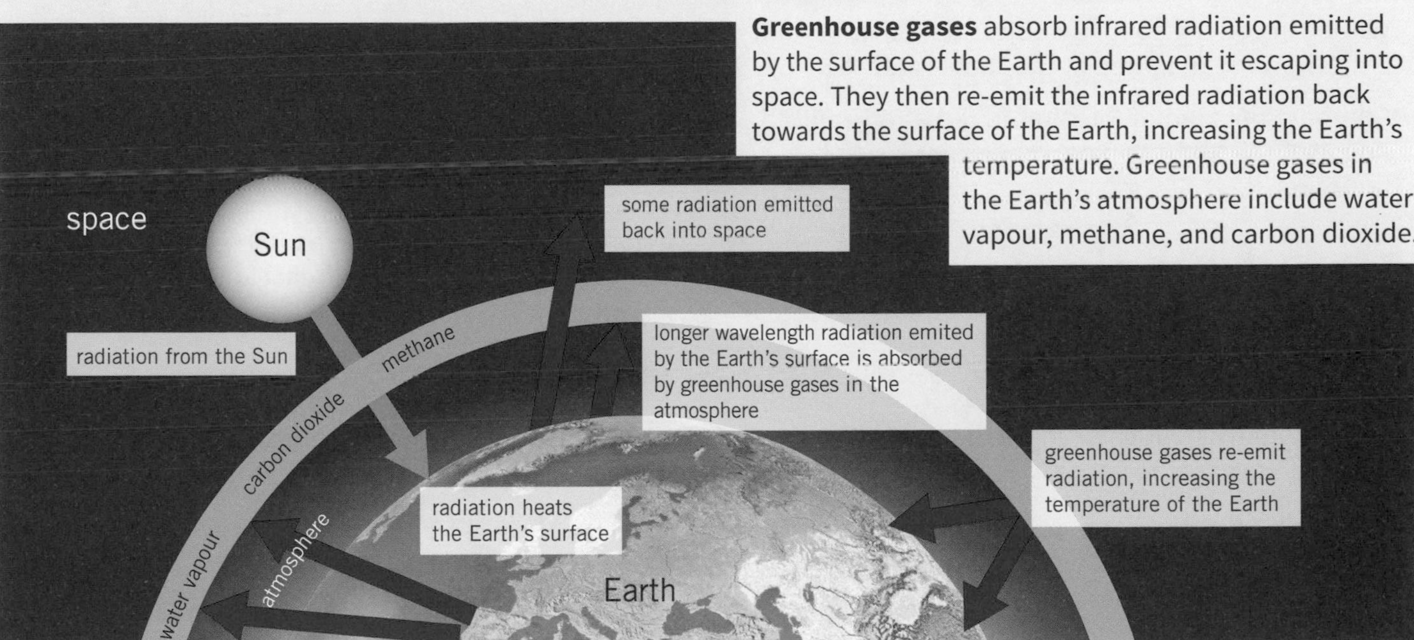

Human activities such as burning fossil fuels, deforestation, and livestock farming are increasing the amount of greenhouse gases in the Earth's atmosphere. This is causing the Earth's temperature to increase – a major cause of climate change.

infrared radiation   specific heat capacity   thermal conductivity   thermal insulator

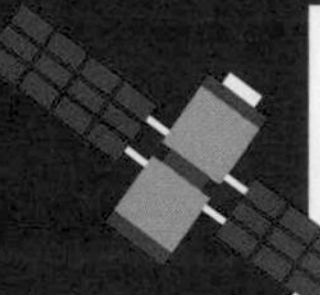

Learn the answers to the questions below then cover the answers column with a piece of paper and write as many as you can. Check and repeat.

	P2 questions		Answers
1	What does a material's thermal conductivity tell you?	Put paper here	how well it conducts heat
2	Which materials have low thermal conductivity?		thermal insulators
3	Give three factors that determine the rate of thermal energy transfer through a material.		thermal conductivity of material, temperature difference, thickness of material
4	What factors affect the rate of heat loss from a building?	Put paper here	thickness of walls and roof, thermal conductivity of walls and roof, the temperature difference between the two sides of the wall/roof
5	Define specific heat capacity.		amount of energy needed to raise the temperature of 1 kg of a material by 1 °C
6	What is infrared radiation?	Put paper here	type of electromagnetic radiation
7	What is the relationship between the temperature of an object and its emission of infrared radiation?		the higher the temperature of an object, the more infrared radiation emitted in a given time
8	What can you tell about an object that absorbs and emits infrared radiation at the same rate?	Put paper here	it is at a constant temperature
9	Compare the amount of infrared radiation emitted and absorbed by an object that is increasing in temperature.		more infrared radiation absorbed than emitted
10	What is a black body?	Put paper here	theoretical object that absorbs 100% of the radiation that falls on it, and does not reflect or transmit any radiation
11	Name three greenhouse gases.		water vapour, carbon dioxide, methane
12	What human activities increase the levels of greenhouse gases released?	Put paper here	(for example) deforestation, burning fossil fuels, livestock farming
13	Why do greenhouse gases increase the Earth's temperature?		Earth's surface absorbs and re-emits radiation from the Sun, which greenhouse gases then absorb – they re-emit this radiation back towards Earth's surface

Now go back and use the questions below to check your knowledge from previous chapters.

Previous questions

Answers

	Previous questions	Answers
1	What is one joule of work?	the work done when a force of 1 N causes an object to move 1 m in the direction of the force
2	Describe the energy transfer when a falling apple hits the ground.	energy is transferred from the kinetic store of the apple and dissipated to the thermal store of the surroundings by sound waves
3	Describe the energy transfer when a moving car slows down.	energy is transferred mechanically from the kinetic store of the car to the thermal store of its brakes – some energy is dissipated to the thermal store of the surroundings
4	Name the unit that represents one joule transferred per second.	watt (W)
5	Describe the energy transfer when an electric kettle is used to heat water.	the electric current in a kettle transfers energy to the heating element's thermal store – energy is then transferred by heating from the heating element's thermal store to the thermal store of the water

Put paper here

Put paper here

Put paper here

Required Practical

Practise answering questions on the required practicals using the example below. You need to be able to apply your skills and knowledge to other practicals too.

Specific heat capacity

To determine changes in specific heat capacity you need to measure mass, temperature rise, and energy transferred (work done).

To do this, you might use an energy meter, or measure time, current, and potential difference (to calculate power).

In the experiment, you need to:

- insulate the block or beaker, use a heatproof mat, and a lid (for a liquid)
- allow the material to heat up before taking measurements (due to thermal inertia)
- add water to make a good thermal contact between the thermometer and a solid material.

Calculated values for specific heat capacity will usually differ from given values because of energy transferred to the surroundings.

Worked example

A student uses a 12 V, 4 A heater to heat a 1 kg metal block. They measure the temperature of the block every minute for 10 minutes.

1 Calculate the work done.

power = potential difference × current
= 12 × 4 = 48 W

work done = power × time
10 min = 600 s
= 48 × 600 = 28 800 J

2 The temperature rise of the block is 75 °C. Calculate the specific heat capacity of the material.

$$\text{specific heat capacity} = \frac{\text{energy transferred}}{\text{mass} \times \text{temperature rise}}$$

$$= \frac{28\,800}{1 \times 75} = 384\ \text{J/kg°C}$$

Practice

A student produces a graph of work done against temperature rise of 0.2 kg of a liquid.

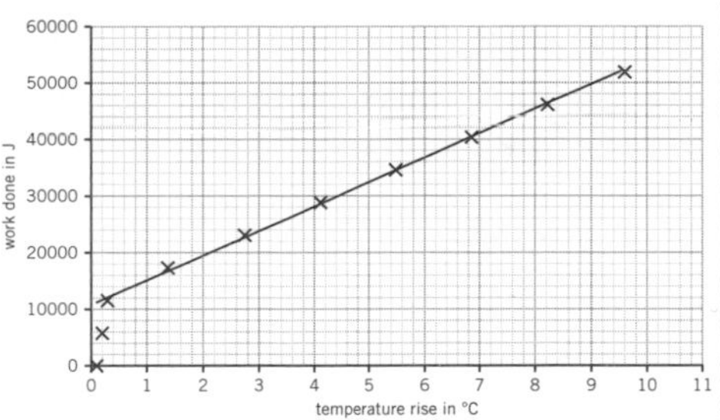

1 Explain why the graph does not go through (0,0).

2 Use the graph to calculate the specific heat capacity. Describe your method.

3 Suggest how you can tell from the graph that the material was well insulated.

Practice

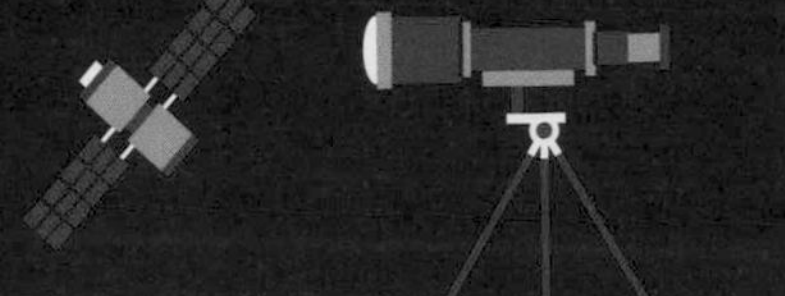

Exam-style questions

01 A block of aluminium has a mass of 1.2 kg.

It is at room temperature, which is 20 °C.

A student uses a heater to increase the temperature to 50 °C.

01.1 Calculate the difference between the initial and final temperatures. **[1 mark]**

____

Temperature change = ______________

01.2 The specific heat capacity of aluminium is 900 J/kg °C.

Calculate the energy transferred to the aluminium to raise its temperature.

Use the correct equation from the *Physics Equations Sheet.* **[2 marks]**

____

____

______________ J

> **Exam Tip**
>
> The first thing you must do is write down the equation.
>
> This is a key skill and you need to get into the habit of always writing that down first.

01.3 A student does this experiment and finds that the energy they need to transfer is bigger than the energy calculated in **01.2**.

Suggest why. **[1 mark]**

____

____

02 Houses have insulation to reduce unwanted energy transfers.

02.1 Describe how to reduce unwanted energy transfers through the roof of a house. **[1 mark]**

____

____

02.2 Older houses have one layer of bricks in their walls.

Newer houses usually have more than one layer of bricks.

Describe the link between the thickness of the layer of bricks and the rate of energy transfer through the bricks. **[1 mark]**

____

____

> **Exam Tip**
>
> There is generally a gap between the layers of brick and this has an effect on the way energy is transferred.

02.3 Owners can fill the gaps between the bricks with insulation.

Write down whether the insulation has a low or a high thermal conductivity.

Explain your answer. **[2 marks]**

Exam Tip

High or low isn't going to be enough to get the marks here. You'll need to give the *why* as well.

03 A swimming pool is heated by the Sun.

A paddling pool next to the swimming pool is also heated by the Sun.

A student notices that the temperature of the paddling pool is higher than the temperature of the swimming pool.

He makes the estimates shown in **Table 1**.

Table 1

	Swimming pool	Paddling pool
energy transferred by the Sun	88 000 MJ	28.8 MJ
temperature of pool	25 °C	28 °C
starting temperature	18 °C	18 °C
specific heat capacity of water	4200 J/kg °C	4200 J/kg °C

Exam Tip

Don't let these big numbers worry you. Just plug the numbers (carefully) into your calculator and you'll be fine!

03.1 Use **Table 1** to find the ratio of the mass of water in the paddling pool to the mass of water in the swimming pool.

Use the correct equation from the *Physics Equations Sheet*.

Use an appropriate number of significant figures. **[6 marks]**

Exam Tip

Put the numbers in first before you rearrange the equation.

03.2 At the end of the day the pool owner puts an identical cover over each pool.

If energy transfer is only through the cover, suggest why:

- the swimming pool might take longer to cool down
- the paddling pool might take longer to cool down. **[2 marks]**

04 A student is comparing the specific heat capacities of two liquids **A** and **B**. Both liquids have the same mass. They use a heater to change the temperature of the liquids, and an energy meter to measure the energy transferred to each liquid by the electric current.

It takes 1.8 kJ of energy to raise the temperature of 10 g of liquid **A** by 50 °C. Calculate the specific heat capacity of liquid **A**. Use the correct equation from the *Physics Equations Sheet*. Liquid **B** has a specific heat capacity that is twice that of liquid **A**.

Suggest **two** differences that the student would observe if they heat the 10 g of liquid **B** using the same heater. Justify your answer. **[6 marks]**

Exam Tip

Pull all the key information out of the text first, for example:

Liquid A

mass =

energy used =

temperature change =

05.1 What quantities do you need to find to work out kinetic energy? Choose **one** answer. **[1 mark]**

mass and speed   mass and time   speed and time

05.2 Car **A** and car **B** are both moving in different ways. Car **A** accelerates under a constant force for three seconds at the start of a race. During the same three seconds car **B** is travelling at a steady speed on a motorway.

Compare:

- the ways energy is stored for each car at the start and end of the three seconds
- the way energy is transferred between those stores. **[6 marks]**

Exam Tip

Break your answer up into two paragraphs – one paragraph for each bullet point mentioned in the question.

05.3 All cars require that you add oil to the engine. Suggest **one** benefit of adding oil to the engine. Use the idea of energy to explain your answer. **[2 marks]**

06 A student is investigating the effect of the thickness of insulation on the rate of energy transfer. She fills a can with hot water. She wraps insulation of different thicknesses around the can. She measures the temperature of the water at the start.

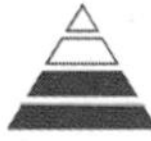

She measures the temperature of the water after 15 minutes.

06.1 Give **one** control variable in this experiment. **[1 mark]**

06.2 The student records the data shown in **Table 2** for the experiment.

Table 2

Thickness of insulation in cm	Temperature at the start in °C	Temperature after 15 minutes in °C
1	75	45
2	75	72
3	75	57
4	75	65

Identify the anomalous result in the data. Suggest a reason why the student may have got this result. **[2 marks]**

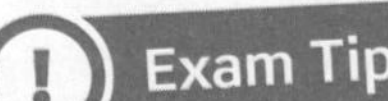

Exam Tip

Suggesting reasons for errors is a key skill.

Read over the method carefully to see if that give you any hints.

06.3 Suggest an improvement to the student's method. Explain the benefit of this improvement. **[2 marks]**

07 A student set up an experiment to measure the specific heat capacity of a 1 kg solid block of an unknown material, as shown in **Figure 1**.

The immersion heater was connected to a power supply.

Figure 1

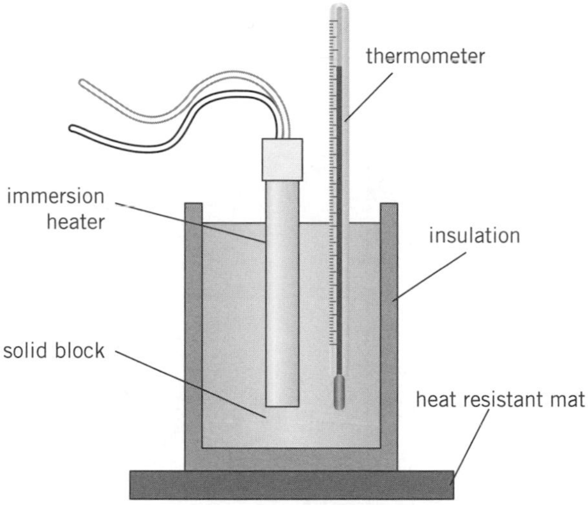

> ### Exam Tip
> You should be familiar with this practical but you may not have seen it drawn like this before.
> Here we show you what is going on inside the practical.

The student measured the starting temperature of the solid block. Then they turned on the power supply and started a stopwatch.

The results are shown in **Table 3**.

Table 3

Time in minutes	Temperature in °C
0	20
2	35
4	45
6	50
8	52

07.1 Plot a graph of temperature against time on **Figure 2**. **[3 marks]**

Figure 2

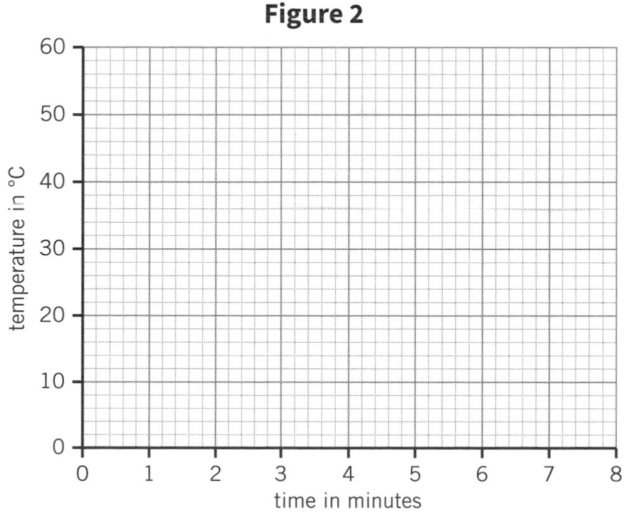

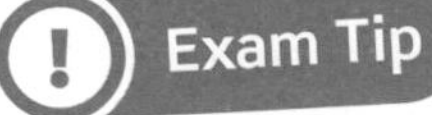

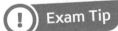

> ### Exam Tip
> Always use crosses to plot points, and draw a line of best fit.

07.2 Describe the change in the rate of temperature increase over time. Explain your answer. **[2 marks]**

Exam Tip

The first step is to write the equation down.

07.3 After 8 minutes, the energy transferred to the block was 15 800 J. Use the data in the table to calculate the specific heat capacity of the block. Use the correct equation from the *Physics Equations Sheet*. Give your answer to three significant figures. **[3 marks]**

07.4 **Table 4** lists the specific heat capacity of some materials.

Table 4

Material	Specific heat capacity in J/kg°C
aluminium	900
iron	452
magnesium	1020
nickel	440
zinc	390

Use **Table 4** to identify which material the block is most likely to be made of. **[1 mark]**

Exam Tip

Use your answer from **07.3**.

08 A student has found a box of different types of insulating materials. They want to work out which type of material is the most effective insulator. They have the following equipment:

- metal cans
- thermometers
- lids
- stopwatches
- insulating materials

08.1 Write down the independent variable in this investigation. **[1 mark]**

08.2 Suggest a dependent variable that they could use to work out the effectiveness of the insulator. **[1 mark]**

Exam Tip

What could we measure and how?

08.3 Describe a method, using the equipment listed, that they could use in order to accurately rank the materials from most to least effective. **[6 marks]**

Exam Tip

When you're planning a practical make sure you give clear instructions that anyone else could follow, and state which equipment you're going to use.

08.4 Suggest a source of uncertainty in this investigation. **[1 mark]**

09 A student connects a temperature sensor to a datalogger inside the classroom. She leaves the sensor and datalogger there overnight. The datalogger produces the graph shown in **Figure 3**.

Figure 3

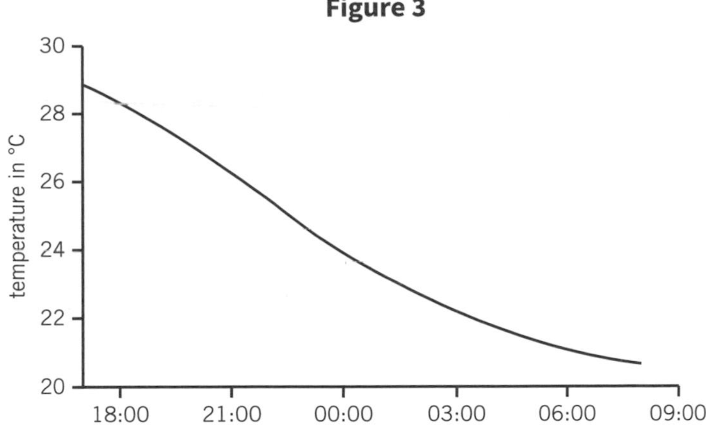

09.1 Explain how you know that the temperature outside the room was lower than the temperature inside the room.
Use ideas of temperature and energy in your answer. **[2 marks]**

Exam Tip

Use data from the graph.

09.2 One feature that affects the rate of cooling is the thickness of the walls of the room.
Give another factor that affects the rate of cooling. **[1 mark]**

09.3 Draw another line on **Figure 3** that shows what would have happened if the walls were thicker.
Explain the position of your line in terms of energy. **[5 marks]**

Exam Tip

Estimate how the drop in temperature would be different.

10 **Table 5** shows data for the thermal conductivities of different materials. Thermal conductivity tells you the rate at which energy is transferred across a thickness of 1 m of the material for each 1 °C temperature difference.

Table 5

Material	Thermal conductivity in W/m °C
A	0.45
B	0.18
C	0.25
D	0.02

10.1 Identify the material (**A**, **B**, **C**, or **D**) with the slowest rate of energy transfer. **[1 mark]**

10.2 A room has walls made of material **A**. The inside of the room is at 25 °C.

The outside of the room is at 15 °C. It takes 2 hours for the room to cool down to the outside temperature. Another room has walls made of material **B**.

Calculate the time it would take for this room to cool down to the outside temperature. Write down **two** assumptions that you have made. **[4 marks]**

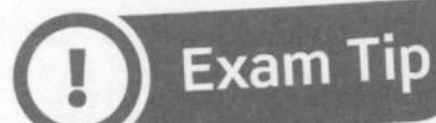

First work out the rate of cooling.

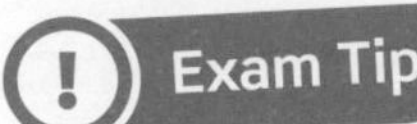

You may not be aware of the assumption you've made here, but it's an important part of science to think about them.

10.3 Thermal conductivity values are not constant. They decrease as the temperature decreases.

Suggest what effect this change will have on the time that you calculated in **10.2**. Give reasons for your answer. **[2 marks]**

11 A teacher shows data from an experiment involving heating oil and water (**Figure 4**).

He wants to compare the liquids in terms of their specific heat capacity.

Figure 4

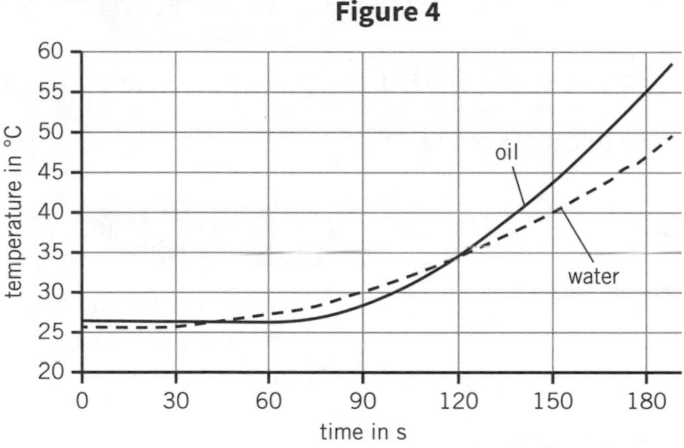

11.1 The teacher does not allow the students to conduct an experiment involving heating oil with a Bunsen burner. Suggest why. **[1 mark]**

11.2 Explain why the heater used to heat the liquids needs to have the same power. **[1 mark]**

Look at the differences in the lines on the graph.

11.3 Compare the relationships between temperature and time for the liquids. **[3 marks]**

11.4 Use the differences between the graphs to compare the specific heat capacity of oil and water.

State any assumptions that you have made. **[4 marks]**

12 A student sees a demonstration involving gallium.

Gallium has a melting point of 29.8 °C.

A small piece of gallium melts in the palm of the demonstrator's hand.

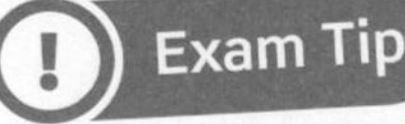

You don't need to know the temperature of the demonstrator's hand, you only need to work out the temperature change.

12.1 Calculate the energy needed to raise gallium to its melting point.

Room temperature = 20 °C.

The specific heat capacity of solid gallium is 371 J/kg °C. **[3 marks]**

12.2 The demonstrator uses a second piece of gallium.

It has three times the mass of the first piece of gallium.

Calculate how much energy would need to be transferred to the second piece to raise it to its melting point. **[2 marks]**

12.3 Aluminium has a greater specific heat capacity than gallium.

Describe what you would notice about the temperature rise of 5 g of aluminium if you transferred the same amount of energy as calculated in **12.1**.

Explain your answer. **[2 marks]**

13 Humans can jump vertically to a height of about 50 cm.

13.1 Write down the equation that links gravitational potential energy, mass, gravitational field strength, and height. **[1 mark]**

> **Exam Tip**
> You can either write this down with words or symbols, just make sure you don't get the symbols confused with other equations.

13.2 Estimate the mass of a human.

Use your estimate to calculate the gravitational potential energy of a human being at a distance of 50 cm off the ground.

Gravitational field strength on Earth = 9.8 N/kg. **[3 marks]**

> **Exam Tip**
> When you estimate the mass of a human, it needs to be a sensible value. For example, 10g would be way too small. But don't spend ages stressing over that bit, as it won't get you any marks.

13.3 The energy to make this jump is stored in the tendons and muscles of the legs. Assume that the muscles and tendons behave like a spring. Assume that the extension of the muscles and tendons is 1 cm.

Calculate the spring constant of the muscles and tendons. Write down **one** assumption that you need to make in doing this calculation. Use the correct equation from the *Physics Equations Sheet*. **[4 marks]**

13.4 A frog jumps vertically to the same height as the human.

The mass of a frog is about 2000 times smaller than that of a human.

Suggest whether the spring constant of the muscles and tendons in a frog's leg is bigger, smaller, or the same as that of the human. Assume that the muscles and tendons stretch by the same amount. Explain your answer. **[4 marks]**

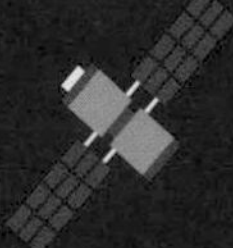

P3 National and global energy resources

Energy resources

The main ways in which we use the Earth's energy resources are:

- generating electricity
- heating
- transport.

Most of our energy currently comes from **fossil fuels** – coal, oil, and natural gas.

Reliability and environmental impact

Some energy resources are more reliable than others. **Reliable** energy resources are ones that are available all the time (or at predictable times) and in sufficient quantities.

Both **renewable** and **non-renewable** energy resources have some kind of **environmental impact** when we use them.

Non-renewable energy resources

- not replaced as quickly as they are used
- will eventually run out

For example, fossil fuels and nuclear fission.

Renewable energy resources

- can be replaced at the same rate as they are used
- will not run out

For example, solar, tidal, wave, wind, geothermal, biofuel, and hydroelectric energies.

	Resource	Main uses	Source	Advantages	Disadvantages
Non-renewable energy resources	coal	generating electricity	extracted from underground	enough available to meet current energy demands reliable – supply can be controlled to meet demand relatively cheap to extract and use	will eventually run out release carbon dioxide when burned – one of the main causes of climate change release other polluting gases, such as sulfur dioxide (from coal and oil) which causes acid rain oil spills in the oceans kill marine life
	oil	generating electricity transport heating			
	natural gas	generating electricity heating			
	nuclear fission	generating electricity	mining naturally occurring elements, such as uranium and plutonium	no polluting gases or greenhouse gases produced enough available to meet current energy demands large amount of energy transferred from a very small mass of fuel reliable – supply can be controlled to meet demand	produces nuclear waste, which is: – dangerous – difficult and expensive to dispose of – stored for centuries before it is safe to dispose of nuclear power plants are expensive to: – build and run – decommission (shut down)

Revision tip

Nuclear is commonly confused as a renewable energy resource as it doesn't release polluting gases - but it's not!

Key terms

Make sure you can write a definition for these key terms.

biofuel · carbon neutral · environmental impact · fossil fuel · geothermal · hydroelectric · non-renewable · reliability · renewable

Renewable energy resources

Resource	Main uses	Source	Advantages	Disadvantages
solar energy	generating electricity	sunlight transfers energy to solar cells	can be used in remote places very cheap to run once installed no pollution/greenhouse gases produced	supply depends on weather expensive to buy and install cannot supply large scale demand
	heating	sunlight transfers energy to solar heating panels		
hydroelectric energy	generating electricity	water flowing downhill turns generators	low running cost no fuel costs reliable and supply can be controlled to meet demand	expensive to build hydroelectric dams flood a large area behind the dam, destroying habitats and resulting in greenhouse gas production from rotting vegetation
tidal energy	generating electricity	turbines on tidal barrages turned by water as the tide comes in and out	predictable supply as there are always tides can produce large amounts of electricity no fuel costs no pollution/greenhouse gases produced	tidal barrages: – change marine habitats and can harm animals – restrict access and can be dangerous for boats – are expensive to build and maintain cannot control supply supply varies depending on time of month
wave energy	generating electricity	floating generators powered by waves moving up and down	low running cost no fuel costs no pollution/greenhouse gases produced	floating generators: – change marine habitats and can harm animals – restrict access and can be dangerous for boats – are expensive to build, install, and maintain dependent on weather cannot supply large scale demand
wind energy	generating electricity	turbines turned by the wind	low running cost no fuel costs no pollution/greenhouse gases produced	supply depends on weather large amounts of land needed to generate enough electricity for large scale demand can produce noise pollution for nearby residents
geothermal energy	generating electricity heating	radioactive substances deep within the Earth transfer heat energy to the surface	low running cost no fuel costs no pollution/greenhouse gases produced	expensive to set up only possible in a few suitable locations around the world
biofuels	generating electricity transport	fuel produced from living or recently living organisms, for example, plants and animal waste	can be **carbon neutral** – the amount of carbon dioxide released when the fuel is burnt is equal to the amount of carbon dioxide absorbed when the fuel is grown reliable and supply can be controlled to meet demand	expensive to produce biofuels growing biofuels requires a lot of land and water that could be used for food production can lead to deforestation – forests are cleared for growing biofuel crops

Retrieval

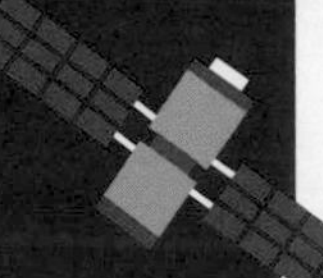

Learn the answers to the questions below then cover the answers column with a piece of paper and write as many as you can. Check and repeat.

	P3 Questions	Answers
1	What is a non-renewable energy resource?	will eventually run out, is not replaced at the same rate it is being used
2	What is a renewable energy resource?	will not run out, it is being (or can be) replaced at the same rate as which it is used
3	What are the main renewable and non-renewable resources available on Earth?	renewable: solar, tidal, wave, wind, geothermal, biofuel, hydroelectric non-renewable: coal, oil, gas, nuclear
4	What are the main advantages of using coal as an energy resource?	enough available to meet current demand, reliable, can control supply to match demand, cheap to extract and use
5	What are the main disadvantages of using coal as an energy resource?	will eventually run out, releases CO_2 which contributes to climate change, releases sulfur dioxide which causes acid rain
6	What are the main advantages of using nuclear fuel as an energy resource?	lot of energy released from a small mass, reliable, can control supply to match demand, enough fuel available to meet current demand, no polluting gases
7	What are the main disadvantages of using nuclear fuel as an energy resource?	waste is dangerous and difficult and expensive to deal with, expensive initial set up, expensive to shut down and to run
8	What are the main advantages of using solar energy?	can be used in remote places, no polluting gases, no waste products, very low running cost
9	What are the main disadvantages of using solar energy?	unreliable, cannot control supply, initial set up expensive, cannot be used on a large scale
10	What are the main advantages of using tidal power?	no polluting gases, no waste products, reliable, can produce large amounts of electricity, low running cost, no fuel costs
11	What are the main disadvantages of using tidal power?	can harm marine habitats, initial set up expensive, cannot increase supply when needed, amount of energy varies on time of month, hazard for boats
12	What are the main advantages of using wave turbines?	no polluting gases produced, no waste products, low running cost, no fuel costs
13	What are the main disadvantages of using wave turbines?	unreliable, dependent on weather, cannot control supply, initial set up expensive, can harm marine habitats, hazard for boats, cannot be used on a large scale
14	What are the main disadvantages of using wind turbines?	unreliable, dependent on weather, cannot control supply, take up lot of space, can produce noise pollution
15	What are the advantages and the disadvantages of using geothermal energy?	advantages: no polluting gases, low running cost disadvantages: initial set up expensive, available in few locations
16	What are the main advantages and disadvantages of using biofuels?	advantages: can be 'carbon neutral', reliable disadvantages: expensive to produce, use land/water that might be needed to grow food
17	What are the main advantages and disadvantages of using hydroelectric power?	advantages: no polluting gases, no waste products, low running cost, no fuel cost, reliable, can be controlled to meet demand disadvantages: initial set up expensive, dams can harm/destroy marine habitats

Put paper here

Now go back and use the questions below to check your knowledge from previous chapters.

Previous questions / Answers

	Previous questions	Answers
1	Name the eight energy stores.	kinetic, gravitational potential, elastic potential, thermal, chemical, nuclear, magnetic, electrostatic
2	What is a system?	an object or group of objects
3	What is work done?	energy transferred when a force moves an object
4	What is infrared radiation?	type of electromagnetic radiation
5	Name three greenhouse gases.	water vapour, carbon dioxide, methane

Put paper here

Put paper here

Required Practical

Practise answering questions on the required practicals using the example below. You need to be able to apply your skills and knowledge to other practicals too.

Thermal insulators

You need to be able to measure the temperature change of a material that has been insulated.

Material type and thickness affect the temperature decrease in a given time, which is known as the rate of cooling.

To be accurate you need to:

- take repeated measurements, and ensure that the starting temperatures are the same
- use the same thickness if you are changing the type of insulator
- use a lid and heatproof mat in all experiments
- take measurements over a long period of time.

Worked example

A student uses two different types of foam to insulate a can of water. Here are their data:

- no insulation: starting temperature 85 °C, final 52 °C
- foam 1: starting temperature 86 °C, final 71 °C
- foam 2: starting temperature 87 °C, final 67 °C

1 Calculate the temperature changes.

85 − 52 = 33 °C

86 − 71 = 15 °C

87 − 67 = 20 °C

2 Name the best insulator. Explain your answer.

Foam 1, as the temperature decrease is the smallest.

3 Suggest why the starting temperature should be the same.

If the water is hotter for one insulator the rate of cooling may be different.

Practice

A student produces a graph of temperature against time for different thicknesses (A, B, C, and D) of an insulator around a beaker of water.

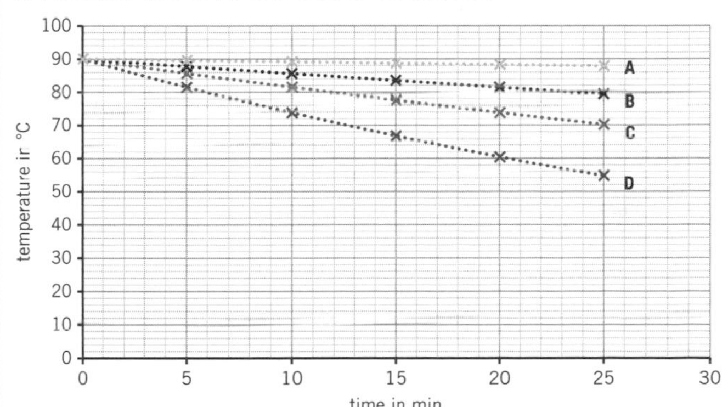

1 Give the letter of the line with the thickest insulator. Explain your answer.

2 Use the graph to calculate the rate of cooling for the thinnest insulator. State the unit.

3 Suggest how the gradients of the graphs would change if the student did not use a lid.

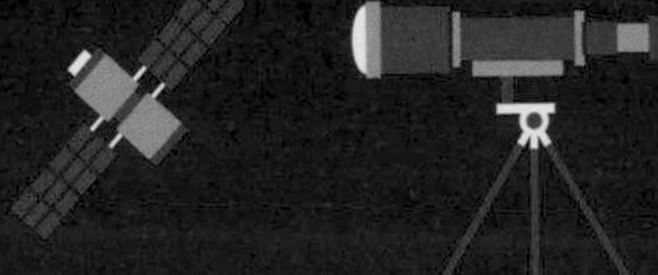

Practice

Exam-style questions

01 Milk is usually kept inside a refrigerator at 4 °C.

A container of milk has been left at room temperature, which is 20 °C.

The mass of the milk is 500 g.

The milk is then put into the refrigerator.

01.1 Calculate the energy transferred from the milk as it cools to 4 °C.

Use an equation from the *Physics Equations Sheet*. **[3 marks]**

The specific heat capacity of milk is 3930 J/kg °C.

____

____

____

______________ J

> **Exam Tip**
> Remember the equation is looking for temperature *change*.

01.2 The doors and walls of the refrigerator are insulated.

Explain in terms of energy why they are insulated. **[2 marks]**

____

____

____

01.3 Suggest whether the insulation material has a high or low thermal conductivity.

Give a reason for your answer. **[2 marks]**

____

____

____

> **Exam Tip**
> Make sure you clearly explain how you came up with your answer.

01.4 Refrigerators have an efficiency rating.

Suggest the link between the efficiency rating and the thermal conductivity.

Explain your suggestions in terms of wasted energy. **[2 marks]**

____

____

____

02 A student finds some data about electricity production and how it has changed over time.

They plot the data on a bar chart, as shown in **Figure 1**.

Figure 1

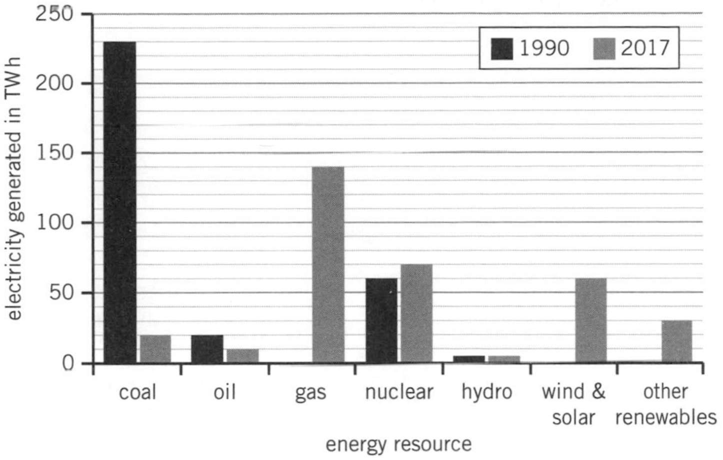

electricity generated in TWh
250
200
150
100
50
0
1990 2017
coal oil gas nuclear hydro wind & solar other renewables
energy resource

02.1 Explain why the student plotted the data as a bar chart and not a line graph. **[2 marks]**

____

____

02.2 Compare the number of renewable resources used in 1990 with the number used in 2017. **[2 marks]**

Use data from the graph.

____

____

____

02.3 Estimate the change in the use of fossil fuels between the two years in terawatt hours (TWh). **[3 marks]**

Don't worry if you've never heard of TWh before. Just use them as the unit for this question, treating them the same as you would any other unit.

____

____

____

____

02.4 There was a large decrease in the use of one non-renewable resource for generating electricity between 1990 and 2017.

Identify this resource.

Suggest why use of this resource in power stations has decreased since 1990. **[2 marks]**

____

____

____

03 A student finds data relating to the cost of generating electricity and the grams of CO_2 produced. A unit is a measure of electricity generation. The CO_2 produced included emissions while the power station was being built and while it is in use. **Table 1** shows the data found by the student.

Table 1

Resource	Cost per unit in p	CO_2 produced per unit in g
solar (the Sun)	40.0	48
nuclear fuel	3.0	12
coal	1.5	820
natural gas	5.0	490
biomass	2.0	230

03.1 Suggest why the cost per unit for solar power is so high. **[1 mark]**

While the sun is free, it's not free to harvest that energy.

03.2 Suggest why the cheapest method in the table may cause environmental problems. Explain your answer. **[3 marks]**

03.3 **Table 1** shows that nuclear fuel is a relatively low-cost option, with the lowest emissions. Suggest **two** disadvantages of using nuclear power stations to generate electricity. **[2 marks]**

Think about safety and renewability.

03.4 The student thinks about how biomass is produced. She suggests that the impact of biomass in terms of CO_2 emissions is actually lower than that shown in **Table 1**. Suggest why. **[3 marks]**

04 A student is investigating how wind turbines work. The student measures the power output of a wind turbine for different wind speeds.

The data for wind speed and power are shown in **Table 2**.

04.1 Plot the data in **Table 2** on a suitable graph. **[5 marks]**

Figure 2

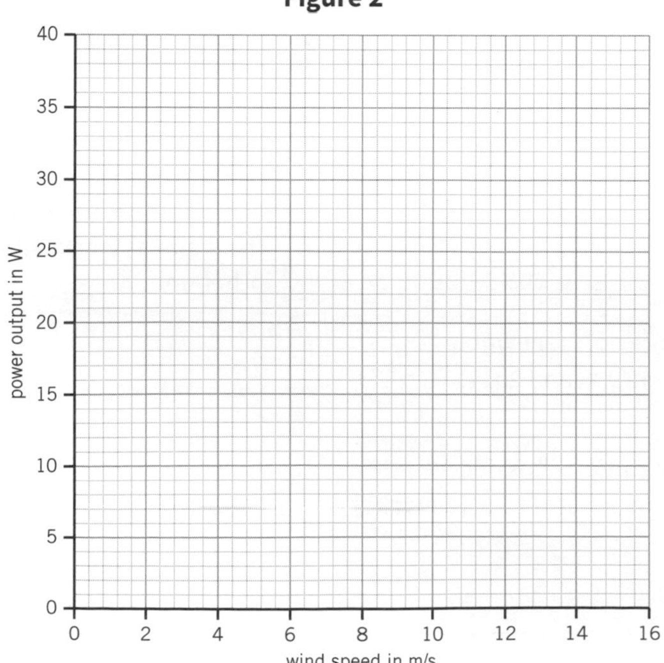

Table 2

Wind speed in m/s	Power output in W
0	0.0
2	0.0
4	0.1
6	1.2
8	2.4
10	3.6
12	3.6
14	3.6

04.2 Describe the relationship between wind speed and power output. **[3 marks]**

04.3 Wind generators can be used to produce electricity used in homes. Suggest **one** advantage and **one** disadvantage of generating electricity using the wind. **[2 marks]**

05 A student collects data from a solar cell. They use a lamp to represent the Sun, and model the effect of clouds by putting sheets of transparent film on top of the solar cell. They make calculations to find the energy per second produced by the solar cell, as given in **Table 3**.

Table 3

Number of sheets of transparent film	Energy per second, test 1	Energy per second, test 2	Mean
0	5.24	5.15	5.20
1	4.12	4.32	4.22
2	3.65	2.11	2.88
3	3.21	3.32	3.27
4	2.50	2.40	2.45
5	1.70	1.60	1.65

05.1 Write down the independent and dependent variables. **[2 marks]**

05.2 List **three** control variables. **[3 marks]**

05.3 Another student looked at the results in **Table 3**. They gave the following feedback.

Statement 1: 'For two sheets, 2.11 is an outlier.'

Statement 2: 'The columns are not labelled correctly.'

Statement 3: 'The significant figures of the measurements are inconsistent.'

Read each of the statements. Write down what action, if any, should be taken as a result. **[3 marks]**

05.4 Calculate the uncertainty in the measurement of energy when one sheet was used. **[2 marks]**

Exam Tip

Independent variables are the ones that you change!

Exam Tip

In your answer make it clear which statements you're referring to. Don't make it hard to mark your work by not laying it out in a neat manner.

06 There are different methods of generating electricity. Some of the resources used to generate electricity are also used for transportation. Some are only used to generate electricity. Some are renewable and some are non-renewable.

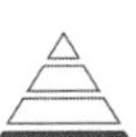

06.1 Describe the difference between a renewable and a non-renewable resource. **[1 mark]**

Exam Tip

The question tells you a lot here. There will be two ticks in each row, and that should help you work out the answer.

06.2 Tick **all** the correct boxes in **Table 4**. There will be a minimum of two ticks in each column. **[3 marks]**

Table 4

Resource	Used to generate electricity	Used as a fuel in cars	Is a renewable resource
coal			
biomass			
oil			
wind			

06.3 Some non-renewable sources cause environmental problems, such as pollution. Describe **one** reason why, despite this fact, they are used to generate electricity. **[1 mark]**

07 The way in which electricity has been generated has changed over time. **Figure 3** shows the changes in the primary sources of energy in the UK between 1990 and 2015.

Figure 3

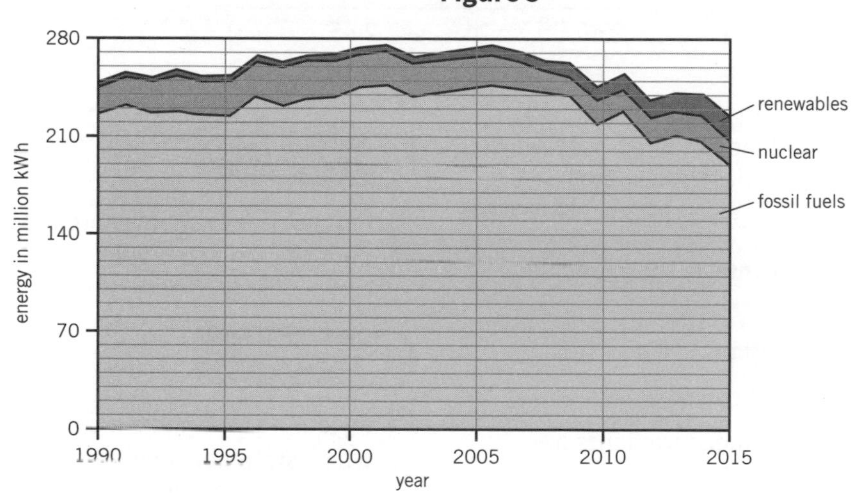

Exam Tip

This may not look like the other graphs you're used to, but it's just a line graph coloured in.

07.1 Name **one** renewable source of energy. **[1 mark]**

07.2 Use data from **Figure 3** to compare the percentage energy use from fossil fuels in 1990 with the percentage use in 2015. Show calculations to justify your answer. **[5 marks]**

Use data from the graph and clearly link it to the statements you make.

07.3 Estimate the rate of decrease of energy use between 2010 and 2015. Use your estimate to determine the number of years it will take for the energy use in the UK to become half of the value in 2015. **[4 marks]**

07.4 Suggest **one** reason why the total energy use might halve in this time and **one** reason why the total energy use might **not** halve in this time. **[2 marks]**

08 A village has a wind turbine installed. The output of the turbine varies between zero and 60 kW. The people in the village are considering increasing their use of renewables using either wind turbines or biomass generators.

They collect the following data:

- annual energy requirement of village – 7 000 MWh
- cost of wind turbine = £1 million
- cost per kWh generated with biofuel (installation and running) = £0.50

Exam Tip

There are lots of non-standard units in this question, so take it carefully step by step.

08.1 Calculate the annual energy requirement of the village in joules (J). **[2 marks]**

08.2 The mean power output of a wind turbine is 33 kW. Calculate how many wind turbines are needed to produce the annual energy requirements of the village. **[4 marks]**

08.3 Calculate which renewable resource would be cheaper to produce the annual energy requirements of the village. **[4 marks]**

Exam Tip

Think about the cost of installation as well as ongoing costs.

08.4 Evaluate the use of each renewable resource in terms of their effect on the environment. **[6 marks]**

09 A student is comparing fossil fuels with energy resources that involve water.

09.1 Name **two** fossil fuels. **[2 marks]**

Exam Tip

There are so many you can pick from – choose the one you know best, not the one you think the examiner wants you to write about.

09.2 The student learns that electricity can be generated using the motion of waves at sea. Name another resource that uses water to generate electricity. Describe how electricity is generated using that resource. **[3 marks]**

09.3 One benefit of using fossil fuels is that they are a reliable resource. Compare the reliability of the resource you described in **09.2** with that of fossil fuels. **[2 marks]**

09.4 Fossil fuels produce carbon dioxide when they burn. Explain why this is an environmental issue. **[2 marks]**

09.5 Describe **one** environmental issue with the resource you described in **09.2**. **[1 mark]**

10 A bungee jumper is standing on a platform attached to a 3.2 m bungee cord. They jump off the platform and touch some water 10 m below. **Figure 4** shows the jumper on the platform.

The mass of the bungee jumper is 60 kg. Gravitational field strength is 9.8 N/kg.

Figure 4

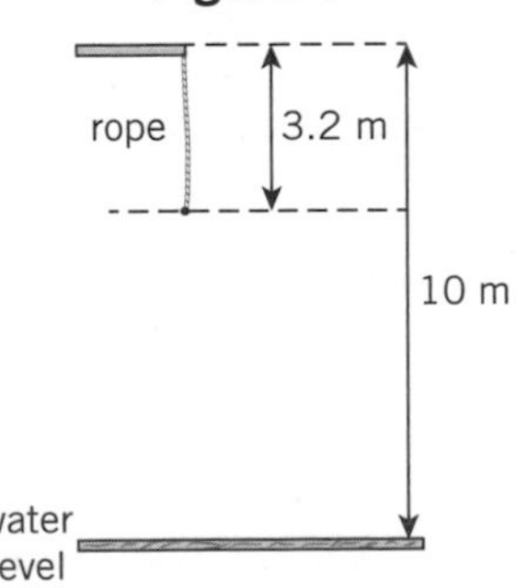

10.1 Write down the equation that links gravitational potential energy, mass, gravitational field strength, and height. **[1 mark]**

10.2 Calculate the gravitational potential energy of the bungee jumper before they jump. **[2 marks]**

10.3 Calculate the spring constant of the bungee cord. Use the correct equation from the *Physics Equations Sheet*. Assume that no energy is wasted during energy transfer. Give your answer to two significant figures. **[5 marks]**

Knowledge

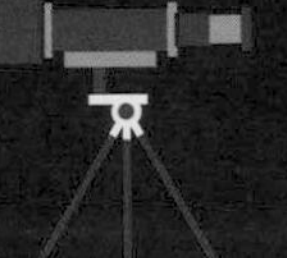

P4 Supplying electricity

Mains electricity

A cell or a battery provides a **direct current (dc)**. The current only flows in one direction and is produced by a **direct potential difference**.

Mains electricity provides an **alternating current (ac)**. The current repeatedly reverses direction and is produced by an **alternating potential difference**.

The positive and negative terminals of an alternating power supply swap over with a regular frequency.

The frequency of the mains electricity supply in the UK is 50 Hz and its voltage is 230 V.

Plugs

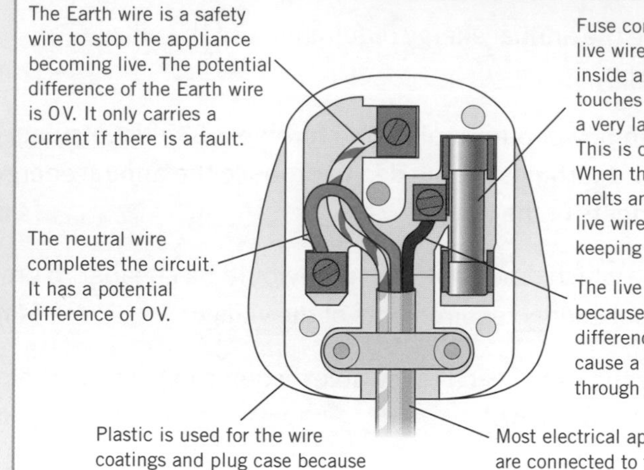

The Earth wire is a safety wire to stop the appliance becoming live. The potential difference of the Earth wire is 0 V. It only carries a current if there is a fault.
Fuse connected to the live wire. If the live wire inside an appliance touches the neutral wire a very large current flows. This is called a **short circuit**. When this happens the fuse melts and disconnects the live wire from the mains, keeping the appliance safe.
The neutral wire completes the circuit. It has a potential difference of 0 V.
The live wire is dangerous because it has a high potential difference of 230 V. This would cause a large current to flow through you if you touched it.
Plastic is used for the wire coatings and plug case because it is a good electrical insulator.
Most electrical appliances in the UK are connected to the mains using a three-core cable. Copper is used for the wires because it is a good electrical conductor and it bends easily.

The National Grid

The **National Grid** is a nationwide network of cables and transformers that link power stations to homes, offices, and other consumers of mains electricity.

Transformers are devices that can change the potential difference of an alternating current.

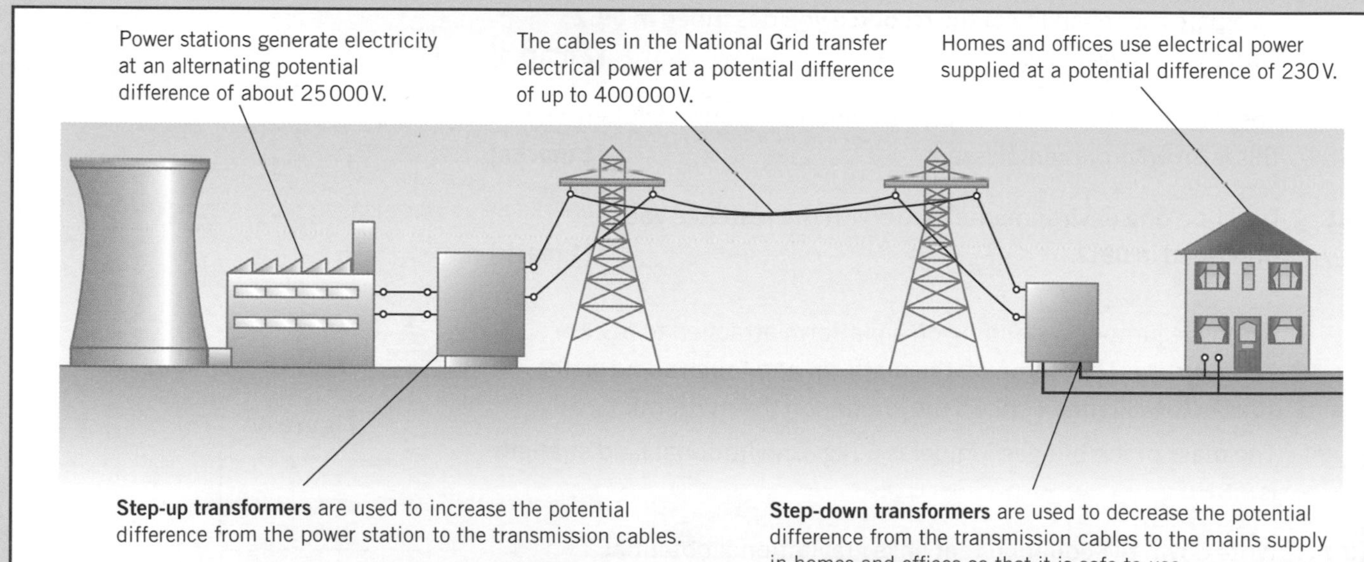

You will learn more about transformers in chapter 18 of this book.

Key terms Make sure you can write a definition for these key terms.

alternating current | alternating potential difference | charge flow | fuse | National Grid | short circuit

Why do transformers improve efficiency?

A high potential difference across the transmission cables means that a lower current is needed to transfer the same amount of power, since:

power (W) = current (A) × potential difference (V)

$$P = IV$$

A lower current in the cables means less electrical power is wasted due to heating of the cables, since the power lost in heating a cable is:

power (W) = current² (A) × resistance (Ω)

$$P = I^2R$$

This makes the National Grid an efficient way to transfer energy.

If 100% efficiency is assumed:

primary potential difference × primary current = secondary potential difference × secondary current

$$V_p I_p = V_s I_s$$

Energy transfer in electrical appliances

Electrical appliances transfer energy.

For example, an hairdryer transfers energy electrically from a chemical store (e.g., the fuel in a power station) to the kinetic energy store of the fan inside the hairdryer and to the thermal energy store of the heating filaments inside the hairdryer.

When you turn an electrical appliance on, the potential difference of the mains supply causes charge (carried by electrons) to flow through it.

You can calculate the **charge flow** using the equation:

charge flow (C) = current (A) × time (s)

$$Q = It$$

You can find the energy transferred to an electrical appliance when charge flows through it using:

energy transferred (J) = charge flow (C) × potential difference (V)

$$E = QV$$

You can find the energy transferred by an electrical appliance using the equation:

power (W) = energy transferred (J) / time (s)

$$P = \frac{E}{t}$$

Revision tip

This topic has the potential for some high level maths questions. If you see a question and you can't decide which equation you need to use to solve it, try looking at combinations of equations.

Revision tip

There are lots of equations in this topic that you need to learn. Find the best way for you to remember them. It could be flashcards, a mnemonic, or changing the lyrics to your favourite song.

coulombs
direct current
direct potential difference
step-down transformer
step-up transformer

Retrieval

Learn the answers to the questions below then cover the answers column with a piece of paper and write as many as you can. Check and repeat.

	P4 questions	Answers
1	Why is the current provided by a cell called a direct current (d.c.)?	only flows in one direction
2	What is an alternating current (a.c.)?	current that repeatedly reverses direction
3	What kind of current is supplied by mains electricity?	alternating current
4	What is the frequency and voltage of mains electricity?	50 Hz, 230 V
5	What colours are the live, neutral, and earth wires in a three-core cable?	live = brown, neutral = blue, earth = green and yellow stripes
6	What is the function of the live wire in a three-core cable?	carries the alternating potential difference from the supply
7	What is the function of the neutral wire in a three-core cable?	completes the circuit
8	What is the function of the earth wire in a three-core cable?	safety wire to stop the appliance becoming live
9	When is there a current in the earth wire?	when there is a fault
10	Why is the live wire dangerous?	provides a large p.d. that would cause a large current to flow through a person if they touched it
11	What is the National Grid?	nationwide network of cables and transformers that link power stations to customers
12	What are step-up transformers used for in the National Grid?	increase the p.d. from the power station to the transmission cables
13	What are step-down transformers used for in the National Grid?	decrease the p.d. from the transmission cables to the mains supply in buildings so that it is safe to use
14	How does having a large potential difference in the transmission cables help to make the National Grid an efficient way to transfer energy?	large p.d. means a small current is needed to transfer the same amount of power, small current in the transmission cables means less electrical power is wasted due to heating
15	What two things does energy transfer to an appliance depend on?	power of appliance, time it is switched on for
16	What are the units for power, current, potential difference, and resistance?	watts (W), amps (A), volts (V), ohms (Ω)

Put paper here

Now go back and use the questions below to check your knowledge from previous chapters.

Previous questions

Answers

	Question	Answer
1	What is a black body?	theoretical object that absorbs 100% of the radiation that falls on it, and does not reflect or transmit any radiation
2	Describe the energy transfer when a ball is fired using an elastic band.	Energy is transferred mechanically from the elastic store of the elastic band to the kinetic store of the band. Some energy is dissipated to the thermal store of the surroundings.
3	What are the main renewable and non-renewable resources available on Earth?	renewable: solar, tidal, wave, wind, geothermal, biofuel, hydroelectric non-renewable: coal, oil, gas, nuclear
4	What are the main advantages and disadvantages of using biofuels?	advantages: can be 'carbon neutral', reliable disadvantages: expensive to produce, use land/water that might be needed to grow food
5	Define specific heat capacity.	amount of energy needed to raise the temperature of 1 kg of a material by 1 °C
6	Name the four ways in which energy can be transferred.	heating, waves, electric current, mechanically (by forces)

Put paper here

Put paper here

Put paper here

Required Practical

Practise answering questions on the required practicals using the example below. You need to be able to apply your skills and knowledge to other practicals too.

Resistance in electrical circuits

You need to be able to measure resistance in an electrical circuit. You can use current and potential difference (p.d.), or an ohmmeter.

Length, cross-sectional area, and material all affect the resistance of a wire.

The arrangement of components affects the resistance of a circuit.

When measuring the resistance of a wire, remember to:

- turn off the power supply when not taking readings to stop the wire getting hot
- fix the wire to a ruler so that the wire is straight
- use crocodile clips that make a good contact with the wire.

When measuring the resistance of a circuit experiment, remember to make sure the ammeter measures the total current.

Worked example

A student uses an ammeter and a voltmeter to measure the resistance of a piece of wire.

Length in cm	10	20	30	40	50
p.d. in V	0.47	0.59	0.64	0.69	0.72
Current in A	0.24	0.16	0.14	0.11	0.10
Resistance in Ω	2.0	3.7	4.6	6.3	

1 Calculate the resistance when the length is 60 cm.

$$\text{resistance} = \frac{\text{p.d.}}{\text{current}} = \frac{0.72}{0.10} = 7.2\ \Omega$$

2 Describe how resistance changes with length of a piece of wire.

As the length of the wire increases, the resistance increases proportionally.

3 Another student finds that resistance does not increase proportionally with the length of wire. Suggest why, and explain your answer.

The wire was still heating up, so the resistance was changing because of temperature not just the change in length.

Practice

Describe how to set up an experiment to compare the resistance of a circuit containing three unequal resistors in parallel with the resistance of a circuit containing three resistors in series.

Include circuit diagrams in your answer.

Practice

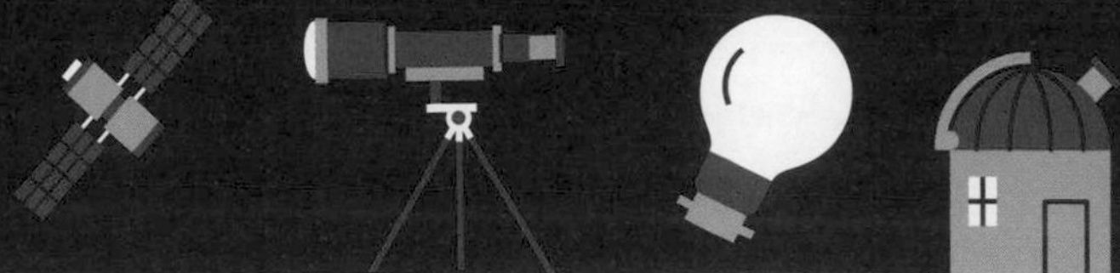

Exam-style questions

01.1 Draw **one** line from each statement beginning to the correct statement ending. You do not need to use all of the endings.

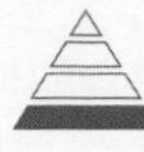

[3 marks]

Statement beginning	Statement ending
The potential difference of the mains electricity in the UK is…	…50 Hz.
	…direct.
The frequency of mains electricity in the UK is…	…about 230 V.
	…100 Hz.
The mains supply in the UK produces a current that is…	…alternating.

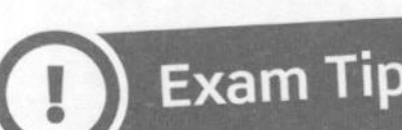

Start this question looking at the units – once you remember the unit for potential difference the answer should become clear.

01.2 Complete the sentences below using the words in the box.

You will need to use some of the words more than once.

live	**earth**	**neutral**

The potential difference between the live and ______________ wires is 230 V.

The potential difference between the ______________ and ______________ wires is 0 V.

When an appliance is connected to the mains and turned on a current flows in the ______________ and ______________ wires.

[5 marks]

01.3 Describe the reason for having an earth wire in a circuit. **[1 mark]**

____

____

____

02 A student has a small electric motor.

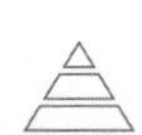

02.1 They connect the motor in a circuit with a 6 V battery.

A current of 1.5 A flows in the circuit.

Show that the power of the motor is 9 W. **[2 marks]**

____

____

____

____

Exam Tip

'Show' questions are great! You already know the answer (9 W), so you just need to clearly show the examiner that you can use an equation to get this answer.

02.2 The student turns the motor on for 30 seconds.

Write down the equation that links power, energy, and time. **[1 mark]**

____

02.3 Calculate the energy transferred by the motor. **[3 marks]**

Exam Tip

For this question you need to use the answer from **02.1**. This is common in exams – you may have to look back at this question to get all the information you need.

______________ J

02.4 The student finds a lamp with the same power rating as the motor.

They connect the lamp to another 6 V battery.

They then turn both circuits on for 30 seconds.

Select the correct statement below. Tick **one** box. **[1 mark]**

Exam Tip

Only tick **one** box – if you tick two you'll get no marks.

The motor transfers more energy than the lamp. ☐

Both devices transfer the same amount of energy. ☐

The lamp transfers more energy than the motor. 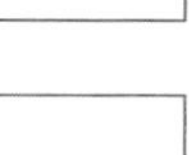

03 **Figure 1** shows how the motor that drives a desk fan is connected to the mains supply.

Figure 1

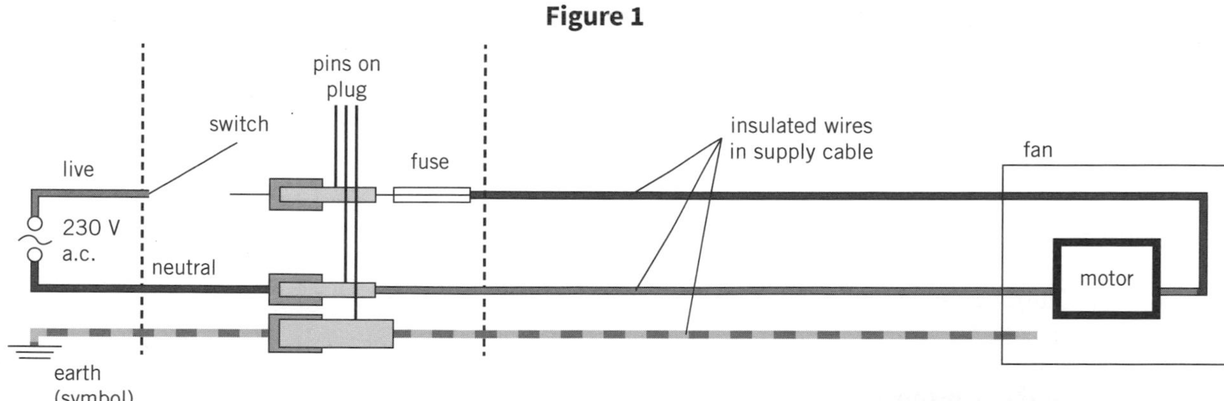

03.1 Use **Figure 1** to explain how the fuse and earth wire prevent a person being injured if there is a fault. **[5 marks]**

03.2 Suggest how to construct the fan so that an earth wire is not required. Explain your suggestion. **[2 marks]**

Exam Tip

This question may seem easy, but look at the number of marks – 5!

To get full marks on this question you must refer back to the information in the figure.

03.3 When the motor is working the current in the wire is 4.5 A.

Calculate the power of the fan motor.

Give your answer to **two** significant figures. **[2 marks]**

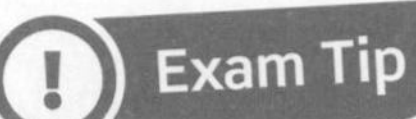

Exam Tip

For this question you'll need to use two different equations to get the final answer. The first clue of this is the fact you're given three numbers in the question.

03.4 There is a fault and the current in the fuse reaches a value of 5.0 A.

The fuse melts in a time of 0.63 seconds.

The energy needed to melt the fuse is 5.4 J.

Calculate the resistance of the fuse wire to an appropriate number of significant figures. **[4 marks]**

04 The National Grid is made up of substations that contain transformers.

04.1 Describe the role of a transformer. **[1 mark]**

04.2 A power station supplies 80 MW to the National Grid.

Some parts of the National Grid operate at very high potential differences.

Compare the power losses of transmitting power using a potential difference of 400 000 V with using a potential difference of 4000 V.

Assume that the wires have a resistance of about 4 Ω.

Use your calculations to explain why it is important to transmit power in the National Grid at a high potential difference rather than low potential difference. **[6 marks]**

Exam Tip

You'll need to do two calculations for this question.

Make it clear which section of information each calculation is referring to in your answer.

05 A student looks at two appliances and produces this table.

Both appliances work when connected to the mains supply.

Table 1

Appliance	Power rating
toaster	1200 W
kettle	2.0 kW

Exam Tip

Watch out for the change between W and kW in this question.

05.1 Write down the equation that links power, potential difference, and current. **[1 mark]**

05.2 Use the potential difference of the mains supply and the data in the table to calculate the current flowing in the wires in the kettle when it is turned on.

Give your answer to **two** significant figures. **[6 marks]**

Exam Tip

There are two ways to approach this question: rearrange the equation and then put the numbers in, or put the numbers in and then rearrange the equation.

In an exam you can get marks for putting the numbers in the right places, so it's a good idea to do that bit first!

05.3 Give the equation that links current, potential difference, and resistance. **[1 mark]**

05.4 Show that the resistance of the kettle is approximately 26 Ω.

Give your answer to **two** significant figures. **[4 marks]**

05.5 Write the equation to calculate energy transfer from power and time. **[1 mark]**

Exam Tip

This answer needs to be given in minutes, so you don't have to convert time into seconds. Be careful if you get a decimal – remember there are 60 seconds in a minute not 100.

05.6 It takes 2 minutes to boil water in the kettle.

Calculate the length of time that the toaster would take to transfer the same amount of energy as the kettle.

Give your answer in minutes. **[6 marks]**

05.7 A student says 'In terms of energy stores and transfers the toaster and kettle are identical.'

Do you agree? Explain your answer. **[2 marks]**

'Yes' or 'no' isn't going to be enough to get full marks on this question. You must explain your reasoning.

06 A student looks at the information on a hair dryer.

The power of the hair dryer is 2000 W.

The potential difference that the hair dryer needs to work is 230 V.

06.1 Write down what 2000 W means in terms of energy and time. **[1 mark]**

06.2 Write down what 230 V means in terms of energy and charge. **[1 mark]**

06.3 The student estimates that it takes 5 minutes to dry their hair.

Write down the equation that links time, power, and energy. **[1 mark]**

Remember that the standard units for time are seconds!

06.4 Calculate the energy transferred from the mains during that time. **[2 marks]**

06.5 Write down the equation that links potential difference, charge, and energy. **[1 mark]**

06.6 Calculate the charge flowing in the hairdryer. Use your answer to **06.4** to help you. **[3 marks]**

07 A student has found a box of metal rods.

The metal rods are numbered but the type of material that each rod is made of is not clear.

The student wants to put the rods in order from best to worst conductor of thermal energy.

They attach a small nail to the rod with wax.

The equipment the student uses is shown in **Figure 2**.

Figure 2

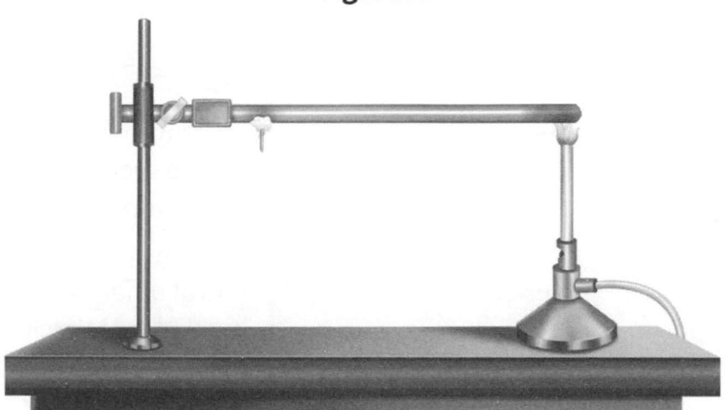

07.1 Design a results table that could be used to collect the data in this experiment. **[3 marks]**

07.2 Write down **two** control variables in the experiment. **[2 marks]**

> **Exam Tip**
>
> Control variables are the ones we keep the same.

07.3 Explain why it would be difficult to collect valid data. **[2 marks]**

07.4 Suggest an improvement to this method that would improve the quality of the data. **[1 mark]**

08 **Table 2** shows a survey of some electrical appliances in a student's house.

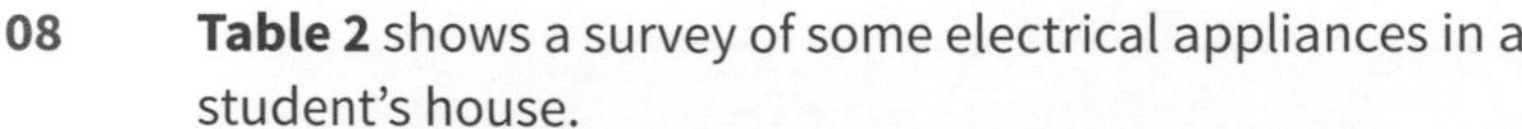

All of the devices use mains p.d.

Table 2

Appliance	Power rating in W	Potential difference in V
hairdryer	2200	
iron	2800	
toaster	2000	

> **Exam Tip**
>
> You may think you need a calculator and equations for this, but this is *not* a maths question. The key word in the question is *mains* p.d.

08.1 Write down the potential difference that should go in the third column.
Explain your answer. **[2 marks]**

08.2 Name the type of energy store that fills when **all** the appliances are being used. **[1 mark]**

08.3 Put the appliances in order from largest to smallest current.
Explain your reasoning. **[2 marks]**

> **Exam Tip**
>
> Correct use of significant figures comes up all the time. This may not be something you've been taught in your science lessons, but I hope you've covered it in maths.
>
> Don't let the fact this is a topic you've covered in maths freak you out!

08.4 The current of the toaster is 8.7 A.
Calculate the resistance using the equation:

$$\text{power} = (\text{current})^2 \times \text{resistance}$$

Give your answer to **two** significant figures. **[4 marks]**

09 A student is using a solar cell.
When they connect it to a voltmeter it produces a potential difference.

09.1 Explain why electricity generated with solar cells is 'renewable'. **[1 mark]**

09.2 The Sun radiates about 4×10^{26} joules per second.
Write down the power of the Sun. **[1 mark]**

> **Exam Tip**
>
> This question may be unlike anything you've seen before, but don't let that worry you!
>
> In the exam you'll get questions on lots of unfamiliar contexts – use a highlighter to pick out the key bits of information and you'll see this questions isn't as hard as it looks!

09.3 Here are some data about the Sun and the energy that is incident on the Earth.
The energy that is incident upon 1 m^2 of the Earth's surface in northern Europe is about 500 J per second.
There are 3.1×10^7 seconds in a year.
Calculate the energy absorbed by a solar cell in northern Europe with an area of 1 m^2 in one year. **[2 marks]**

09.4 The world demand for electricity is about 7×10^{18} J per year.

Calculate the area of solar cells placed in northern Europe that would be needed to meet the world demand for energy. **[2 marks]**

10 A trampoline is made of a rubber sheet attached to a frame by springs.

Each spring stretches by 0.01 m when a person stands on the trampoline.

The spring constant of each spring is 500 N/m.

10.1 Calculate the energy stored in the spring. Use the correct equation from the *Physics Equations Sheet*. **[2 marks]**

10.2 A student bounces on the trampoline.

On the first bounce they reach a height of 2 m above the trampoline.

On the second bounce they reach a height of 1.5 m.

Describe the energy stores at the top of the first bounce and at the top of the second bounce.

Use your description to explain why the second bounce is lower than the first. **[3 marks]**

> **Exam Tip**
>
> Think about wasted energy when answering this question.

10.3 Describe **two** processes by which the energy is transferred between the energy stores in **04.2**. **[2 marks]**

11 A student is explaining how electricity gets to their house.

They draw a diagram to show this (**Figure 3**).

Figure 3

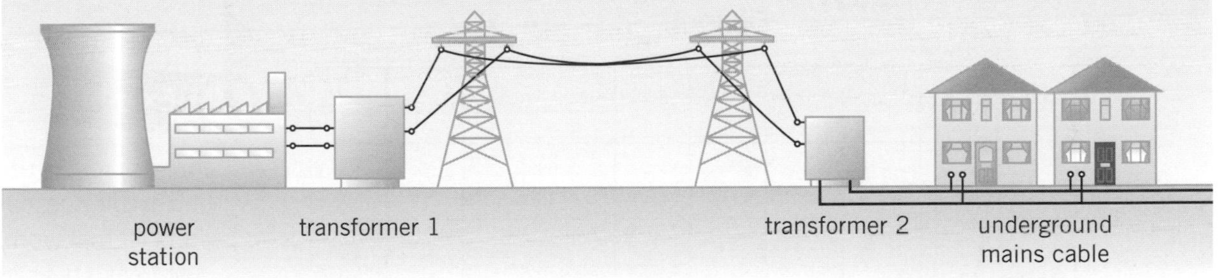

11.1 Write down the name of the system of transformers and power cables that provides electricity to businesses and houses. **[1 mark]**

> **Exam Tip**
>
> For compare questions you have to give ways they are the same and ways they are different.

11.2 A younger student asks:
'Why are there two transformers? Are they both the same?'

Answer these questions by comparing the transformers. **[3 marks]**

11.3 Give **one** reason why the system is an efficient way of transferring energy. **[2 marks]**

> **Exam Tip**
>
> It's important to look at the number of marks available for question.
>
> This question asks for **one** reason but has **two** marks available. You can assume there is one mark available for your reason and one for why.

12 A student learns that the oven in their kitchen works on a separate circuit to that of the toaster and other small electrical appliances.

They find the following information:

- the oven has a power of 9 kW
- the toaster has a power of 2 kW
- ovens are connected to the mains with much thicker wires than other appliances.

12.1 Calculate the current flowing in the oven. **[3 marks]**

12.2 Compare the current flowing in the oven with that of the toaster. **[4 marks]**

12.3 Suggest why the oven is connected with thicker wires. **[2 marks]**

12.4 Current larger than 0.1 A is dangerous to the human body. Explain why the student can use both appliances safely. **[2 marks]**

> **Exam Tip**
>
> For **12.1** and **12.2**, you'll need to do two separate calculations and then compare them. Remember to make it very clear which calculation applies to which appliance, and then give the similarities and differences.

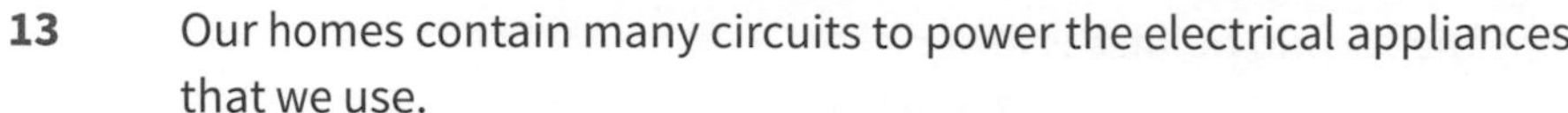

13 Our homes contain many circuits to power the electrical appliances that we use.

One example is a circuit that controls the lights in the house – as it gets dark outside, the lights inside come on.

A company making this circuit has the choice of two types of light-dependent resistor (LDR).

The graph in **Figure 4** shows how the resistance of each LDR varies with light intensity.

Light intensity is measured in lux.

Figure 4

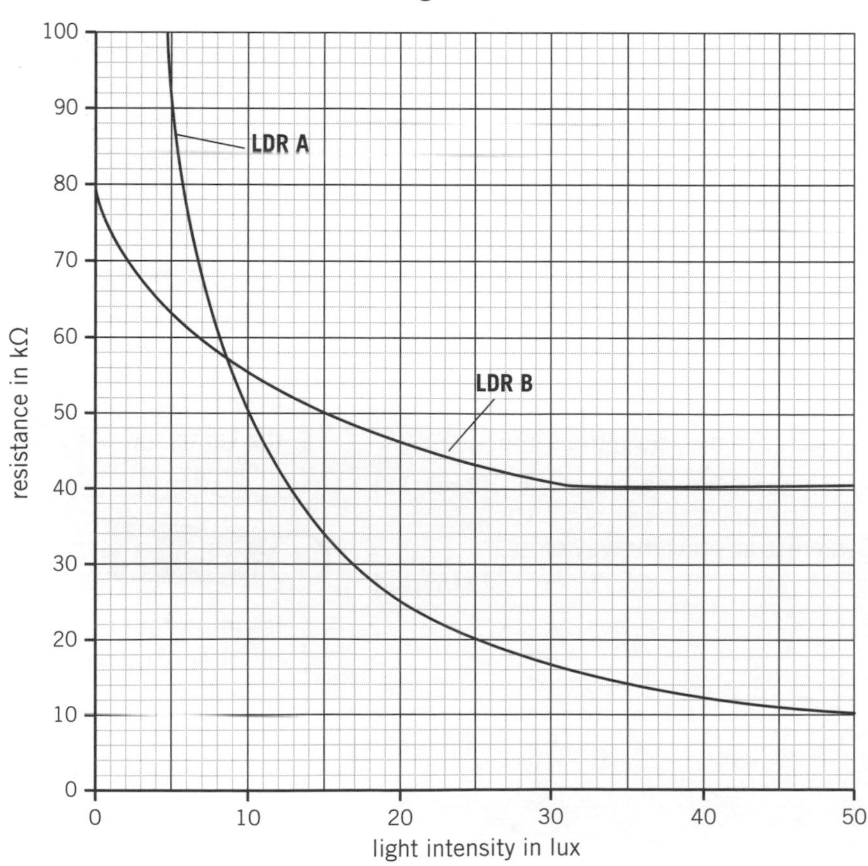

> **Exam Tip**
>
> You may not be used to the unit in this question, but don't let that worry you. Just treat it the same way you would any other unit.

13.1 Describe the relationship between resistance and light intensity for LDR **A**. **[2 marks]**

13.2 Compare the relationship between resistance and light intensity for LDR **A** with this relationship for LDR **B**. **[3 marks]**

13.3 The company sets up a test circuit for LDR **A** (**Figure 5**).

Figure 5

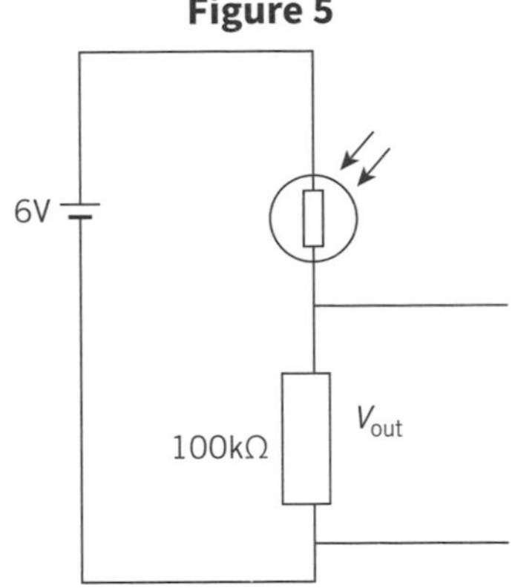

Calculate the potential difference across the resistor (V_{out}) when the light intensity is 13 lux. **[4 marks]**

13.4 The potential difference, V_{out}, is used to turn on lights inside the house when the light level outside is low.

The sensor needs a value of V_{out} larger than 5 V to turn the lights on.

Suggest which LDR the company should use to achieve a V_{out} of 5 V.

Explain your answer. **[6 marks]**

14 A student is comparing the brightness of two bulbs connected in different types of circuit.

Bulb **A** has a resistance of 5 Ω.

Bulb **B** has a resistance of 10 Ω.

They connect the two bulbs in series with a 12 V battery.

Then they connect the same two bulbs in parallel with the same 12 V battery.

Compare the brightness of the bulbs in the series and parallel circuits.

Justify your answer with calculations. **[6 marks]**

> **Exam Tip**
>
> Tell the story of the line, use data from the graph, and describe the pattern.

> **Exam Tip**
>
> You'll need to get the resistance from the graph to help with this question – find 13 lux on the x-axis and read up.

> **Exam Tip**
>
> This question has many parts.
>
> Step 1 – calculate the resistance a V_{out} of 5 V would give
>
> Step 2 – look at the graph to see which LDR works at this resistance
>
> Step 3 – explain your answer, writing down why

> **Exam Tip**
>
> Brightness is dependent on power, which is current × voltage.
>
> In a series circuit the current is the same everywhere, but not in a parallel circuit.

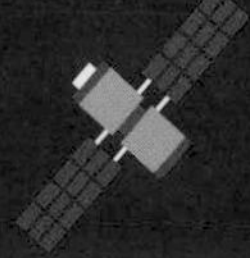

P5 Electric circuits

Charge

An atom has no **charge** because it has equal numbers of positive protons and negative electrons.

When electrons are removed from an atom it becomes *positively* charged. When electrons are added to an atom it becomes *negatively* charged.

Static charge

Insulating materials can become charged when they are rubbed with another insulating material. This is because electrons are transferred from one material to the other. Materials that gain electrons become negatively charged and those that lose electrons become positively charged.

Positive charges do not usually transfer between materials.

Electric charge is measured in coulombs C.

Sparks

If two objects have a very strong electric field between them, electrons in the air molecules will be strongly attracted towards the positively charged object. If the electric field is strong enough, electrons will be pulled away from the air molecules and cause a flow of electrons between the two objects – this is a **spark**.

Electric fields

A charged object creates an **electric field** around itself.

If a charged object is placed in the electric field of another charged object it experiences **electrostatic force**. This means that the two charged objects exert a non-contact force on each other:

- like charges repel each other
- opposing charges attract each other.

The electric field, and the force between two charged objects, gets stronger as the distance between the objects decreases.

Drawing electric fields

Electric fields can be represented using a diagram with field lines. These show the direction of the force that a small positive charge would experience when placed in the electric field.

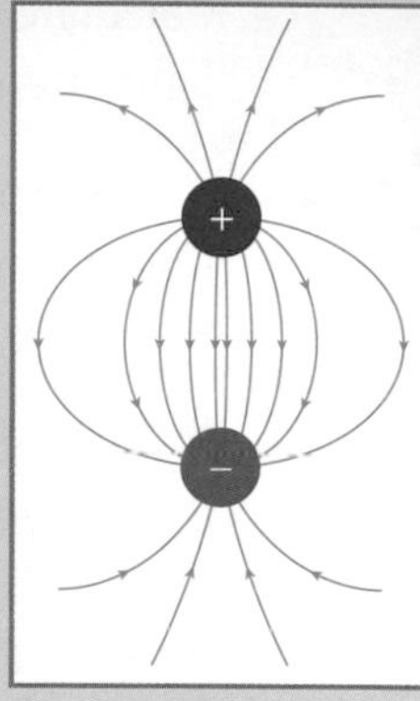

When drawing electric fields, make sure:

- field lines meet the surface of charged objects at 90°
- arrows always point away from positive charges and towards negative charges.

Electric current

Electric current is when **charge** flows. The charge in an electric circuit is carried by electrons.

The unit of current is the ampere (amp, A).

1 ampere = 1 coulomb of charge flow per second

In circuit diagrams, current flows from the positive terminal of a cell or battery to the negative terminal. This is known as conventional current.

In a single closed loop, the current has the same value at any point in the circuit.

Metals are good conductors of electricity because they contain delocalised electrons, which are free to flow through the structure.

Potential difference

Potential difference (p.d.) is a measure of how much energy is transferred between two points in a circuit. The unit of potential difference is the volt (V).

- The p.d. across a component is the work done on it by each coulomb of charge that passes through it.
- The p.d. across a power supply or battery is the energy transferred to each coulomb of charge that passes through it.

For electrical charge to flow through a circuit there must be a source of potential difference.

Key terms

Make sure you can write a definition for these key terms.

ampere
charge
coulomb
current
electric field
electrostatic force
parallel
potential difference
resistance
series
static

Resistance

When electrons move through a circuit, they collide with the ions and atoms of the wires and components in the circuit. This causes **resistance** to the flow of charge.

The unit of resistance is the ohm (Ω).

A long wire has more resistance than a short wire because electrons collide with more ions as they pass through a longer wire.

The resistance of an electrical component can be found by measuring the current and potential difference:

potential difference (V) = current (A) × resistance (Ω)

$V = IR$

Current-potential difference graphs

A graph of current through a component against the p.d. across it (I–V graph), is known as the component characteristic.

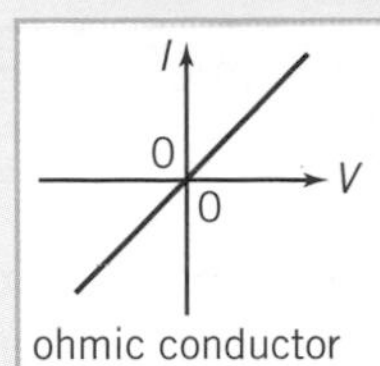

Current is directly proportional to the p.d. in an ohmic conductor at a constant temperature. The resistance is constant.

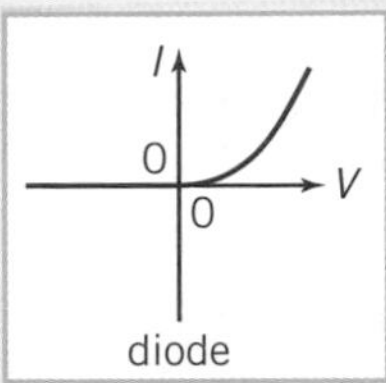

The current through a diode only flows in one direction – called the forward direction. There needs to be a minimum voltage before any current will flow.

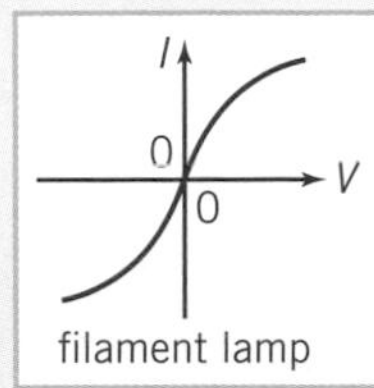

As more current flows through the filament, its temperature increases. The atoms in the wire vibrate more, and collide more often with electrons flowing through it, so resistance increases as temperature increases.

The resistance of an ohmic conductor can be found by calculating the gradient at that point and taking the inverse:

$$\text{resistance} = \frac{1}{\text{gradient}}.$$

Circuit components

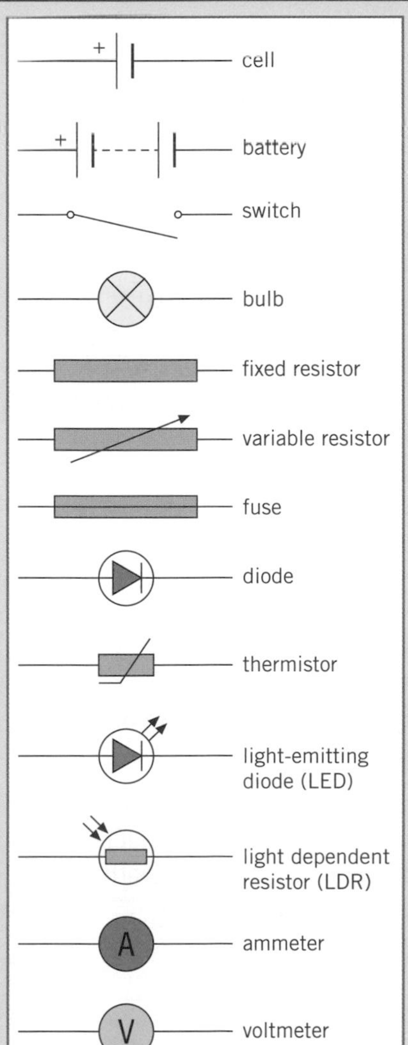

Series circuits

In a series circuit, the components are connected one after the other in a single loop. If one component in a series circuit stops working the whole circuit will stop working.

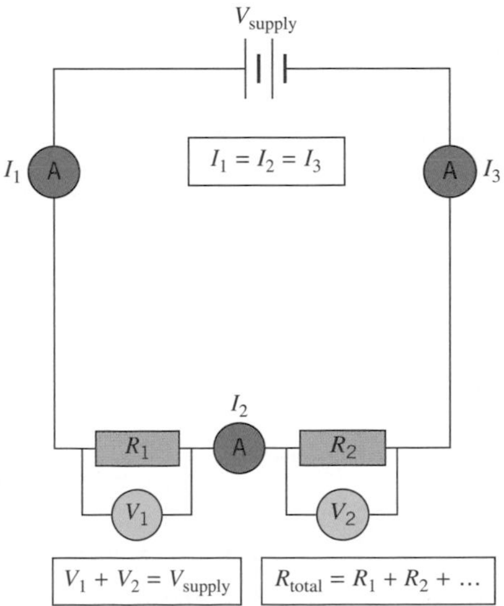

Components with a higher resistance will transfer a larger share of the total p.d. because $V = IR$ (and current is the same through all components).

Parallel circuits

A parallel circuit is made up of two or more loops through which current can flow. If one branch of a parallel circuit stops working, the other branches will not be affected.

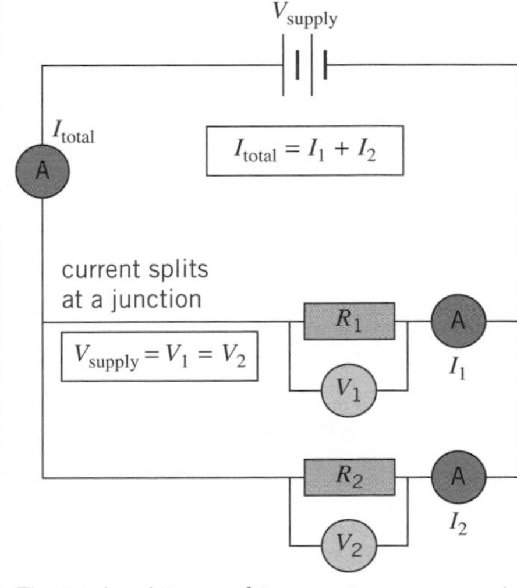

The total resistance of two or more components in parallel is always less than the smallest resistance of any branch. This is because adding a loop to the circuit provides another route for the current to flow, so more current can flow in total even though the p.d. has not changed. Adding more resistors in parallel decreases the total resistance of a circuit.

Retrieval

Learn the answers to the questions below then cover the answers column with a piece of paper and write as many as you can. Check and repeat.

	P5 questions	Answers
1	How does a material become charged?	becomes negatively charged by gaining electrons and becomes positively charged by losing electrons
2	What will two objects carrying the same type of charge do if they are brought close to each other?	repel each other
3	What is an electric field?	region of space around a charged object in which another charged object will experience an electrostatic force
4	What happens to the strength of an electric field as you get further from the charged object?	it decreases
5	What is electric current?	rate of flow of charge
6	What units are charge, current, and time measured in?	coulombs (C), amperes (A), seconds (s) respectively
7	What is the same at all points when charge flows in a closed loop?	current
8	What must there be in a closed circuit so that electrical charge can flow?	source of potential difference (p.d.)
9	Which two factors does current depend on and what are their units?	resistance in ohms (Ω), p.d. in volts (V)
10	What happens to the current if the resistance is increased but the p.d. stays the same?	current decreases
11	What is an ohmic conductor?	conductor where current is directly proportional to the voltage so resistance is constant (at constant temperature)
12	What happens to the resistance of a filament lamp as its temperature increases?	resistance increases
13	What happens to the resistance of a thermistor as its temperature increases?	resistance decreases
14	What happens to the resistance of a light-dependent resistor when light intensity increases?	resistance decreases
15	What are the main features of a series circuit?	same current through each component, total p.d. of power supply is shared between components, total resistance of all components is the sum of the resistance of each component
16	What are the main features of a parallel circuit?	p.d. across each branch is the same, total current through circuit is the sum of the currents in each branch – total resistance of all resistors is less than the resistance of the smallest individual resistor

Put paper here

Now go back and use the questions below to check your knowledge from previous chapters.

Previous questions | Answers

	Previous questions	Answers
1	What units are charge, current, and time measured in?	coulombs (C), amperes (A), seconds (s) respectively
2	What is an electric field?	region of space around a charged object in which another charged object will experience an electrostatic force
3	What happens to the resistance of a light-dependent resistor when light intensity increases?	resistance decreases

Put paper here

Required Practical

Practise answering questions on the required practicals using the example below. You need to be able to apply your skills and knowledge to other practicals too.

I–V graphs

You need to be able to determine the relationship between current and potential difference (p.d.) for a lamp, resistor, and diode.

You should be able to draw and interpret *I–V* graphs – the shape of the graph is characteristic of the component. The gradient of the graph is not related to the resistance, but resistance can be calculated from values for p.d. and current.

A variable power supply or resistor should be used to change the current in both directions.

The diode needs to be connected with a protective resistor so it does not get too hot.

Worked example

A student uses a circuit to measure values of current and p.d. for component **X**. She decides to use a variable resistor to change the current.

1 Below are some data for the component. Plot a graph of the data and draw a line of best fit.

p.d. in V	0.1	0.4	0.6	0.8	1
Current in A	0.38	0.59	0.64	0.69	0.72
p.d. in V	−0.1	−0.4	−0.6	−0.8	−1
Current in A	−0.38	−0.59	−0.64	−0.69	−0.72

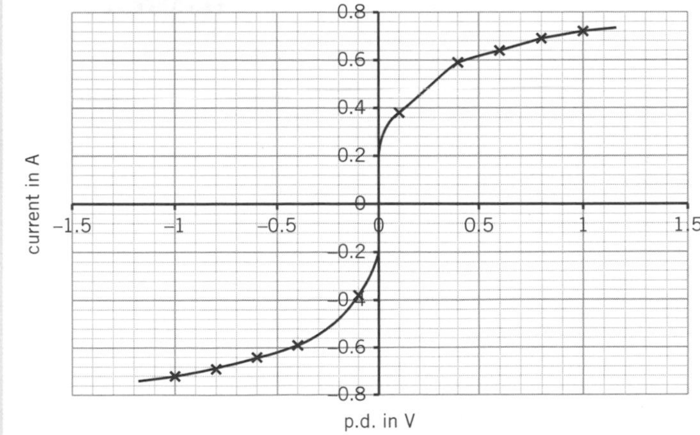

2 Suggest the name of component **X**. Use the graph and data to explain your answer.

X is a lamp as the graph is symmetrical:

$$\text{resistance} = \frac{\text{p.d.}}{\text{current}}$$

$$R = \frac{0.1}{0.38} = 0.26\,\Omega \qquad R = \frac{1}{0.72} = 1.39\,\Omega$$

So as p.d. increases, resistance increases.

Practice

A student has set up an experiment to collect data to plot an *I–V* graph for a piece of resistance wire.

1 Give two changes the student will need to make to repeat the experiment with a diode.

2 Sketch the *I–V* graph for a diode. Explain the shape of the graph in terms of resistance.

Exam-style questions

01 A student finds a box of different resistors. One of the resistors is not marked.

The student wants to find the resistance. They place the unknown resistor in the circuit shown in **Figure 1**.

Figure 1

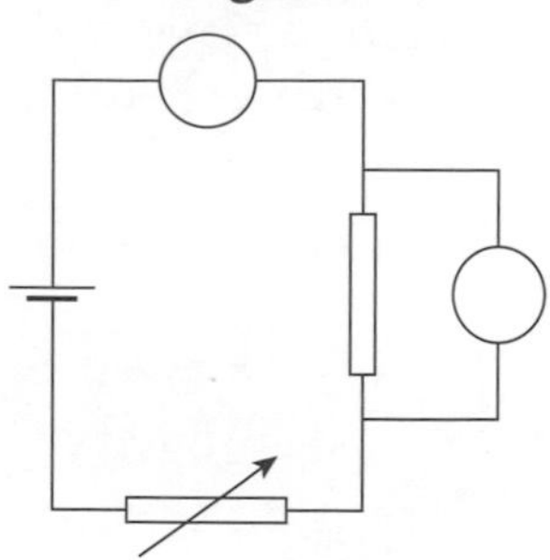

01.1 The student has not labelled the diagram.

The ammeter and voltmeter need to be in the correct places.

Write the letters **A** and **V** in the circles above to show where to put the meters.

Label the resistor.

Label the variable resistor. **[4 marks]**

01.2 The student measures a current of 0.3 A.

Define current. **[1 mark]**

____

> **Exam Tip**
>
> Look back at the Knowledge page – this is practice, not the real exam, so you are allowed to look back!

01.3 Write down the equation that links current, potential difference, and resistance. **[1 mark]**

____

01.4 The student measures a potential difference of 6 V.

Calculate the resistance of the resistor.

____

____

____

____

Resistance = ______________

[3 marks]

> **Exam Tip**
>
> For all maths questions:
>
> Step 1 – write down the equation you are using (part **01.3**)
>
> Step 2 – put the numbers into the equation
>
> Step 3 – rearrange the equation
>
> Step 4 – do the maths
>
> Step 5 – write down the answer **with units**

02 **Figure 2** shows four circuits drawn by a student.

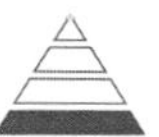

Figure 2

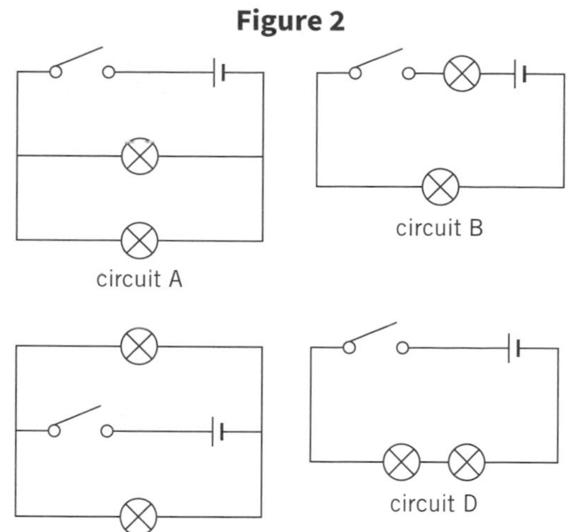

02.1 Here are some statements about the circuits in **Figure 2**.

Tick **one** box. **[1 mark]**

Circuit A is a series circuit	☐
Circuits A and C are parallel circuits	☐
Only circuit A is a parallel circuit	☐
Circuits C and D are series circuits	☐

> **! Exam Tip**
>
> Only tick **one** box.
>
> Ticking two will mean you get no marks!

02.2 Complete these sentences about the three circuits.

Use the letters **A**, **B**, **C**, or **D**. **[3 marks]**

The bulbs in circuits ____________ and ____________ are the brightest.

If one of the bulbs in circuit ____________ or ____________ breaks, the other bulb will go out.

An ammeter placed anywhere in circuit ____________ or ____________ will measure the same current.

02.3 The student looks at circuit **B** and says:

'I think that when you press the switch, the bulb nearer the battery will be brighter than the bulb that is further away.'

Do you agree? Explain your answer. **[3 marks]**

____

____

____

____

> **! Exam Tip**
>
> Get into the habit of looking at the number of marks available, this will help you work out what the examiners are looking for.
>
> This is a 3 mark question, meaning a 'yes' or 'no' answer won't be enough – you need to explain why as well.

03 A student wants to demonstrate the difference between the equivalent resistances of two resistors in series and parallel circuits.

Describe how a student could set up two circuits, one in series and one in parallel, to show the difference.

You need to:

- draw a circuit diagram for each circuit
- describe the measurements the student should make, and how the student should use those measurements to calculate the equivalent resistance of each circuit
- describe the differences in the equivalent resistance that the student should find. **[6 marks]**

Always use a ruler for circuit diagrams!

Exam Tip

Think about which equations you'll need to calculate resistance and which circuit components can give you the information you'll need.

04 A student is using a thermistor and a data logger to monitor the change in temperature in the school greenhouse.

04.1 Describe how the resistance of a thermistor depends on temperature. **[1 mark]**

Exam Tip

This is a 'Describe' question, so you need to say *what* will happen but not *why*.

04.2 The student connects the thermistor in the circuit shown in **Figure 3**.

They think that if the temperature changes the reading on the voltmeter will change.

Explain why this would **not** happen. **[2 marks]**

Figure 3

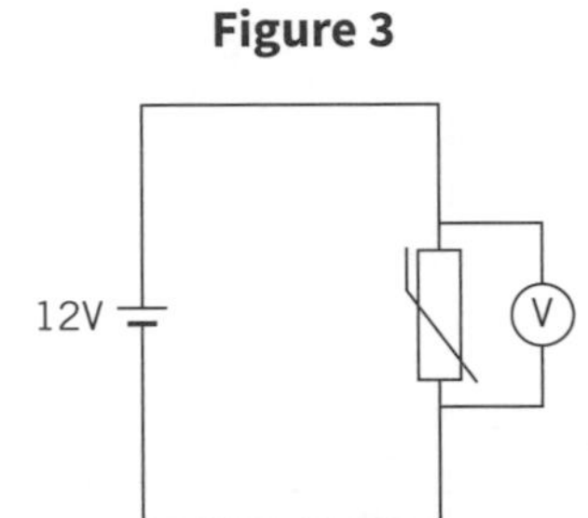

Exam Tip

This question isn't as hard as it looks – just think logically and use your maths skills here!

04.3 Another student sets up the circuit shown in **Figure 4**.

They measure the highest and lowest voltmeter readings.

The highest reading is 8 V. The lowest reading is 3 V.

Use proportion to calculate the resistance of the thermistor when it is very hot, and when it is very cold.

Explain your method. **[7 marks]**

Figure 4

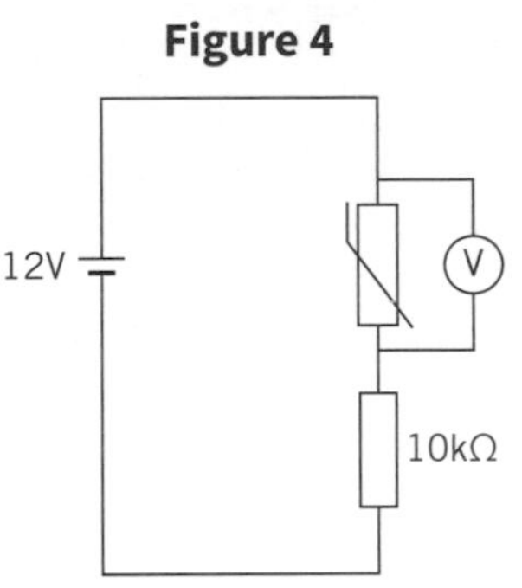

05 **Figure 5** shows a lamp and a resistor connected in series.

Figure 5

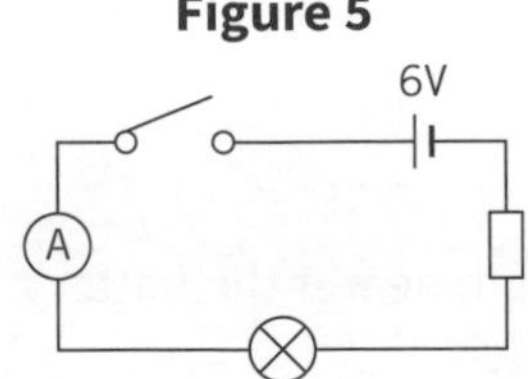

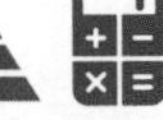

05.1 The potential difference across the lamp is 4 V.

Calculate the potential difference across the resistor. **[2 marks]**

05.2 The reading on the ammeter is 0.2 A.

Calculate the resistance of the lamp. **[3 marks]**

Exam Tip

To answer this you'll need to use your answer from **05.1** (p.d.) and the information in **05.2** (0.2 A). Show questions are great, because you know what the answer is, and if you write your working out clearly you know you've got the marks!

05.3 Show that the resistance of the resistor is 10 Ω. **[2 marks]**

05.4 The student now connects another identical resistor in series with the resistor and lamp in the circuit.

Select the correct word to complete this sentence:

The ammeter reading will **increase / stay the same / decrease**.

Explain your answer. **[2 marks]**

06 The graph in **Figure 6** shows the current vs potential difference graph for two components: **A** and **B**.

Figure 6

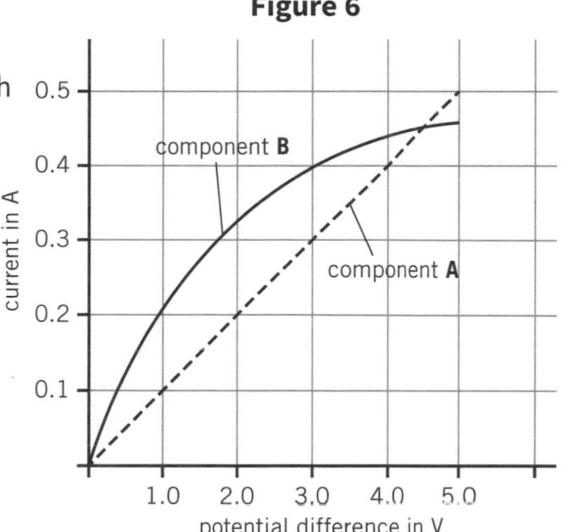

06.1 Write down which component is ohmic. **[1 mark]**

06.2 Calculate the resistance of component **A**. **[3 marks]**

Exam Tip

Find the gradient of the graph!

06.3 Describe what happens to the resistance of each component as the potential difference across it increases.

Explain how you used the graph to work out your answer. **[4 marks]**

Exam Tip

Use data from the graph.

06.4 A student connects the two components in parallel across a 3 V battery.

Calculate the total current in the circuit. **[2 marks]**

06.5 Calculate the resistance of the circuit as set up in **14.4**.

Give your answer to **two** significant figures. **[2 marks]**

07 **Figure 7** shows a sphere that has been charged.

Figure 7

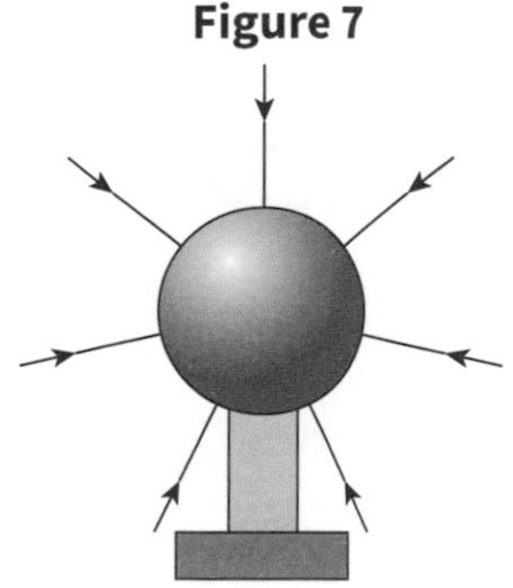

07.1 Write down the charge on the sphere. **[1 mark]**

07.2 A student brings a negatively charged object near the sphere.

Draw an arrow on the object in **Figure 8** to show the force on the object. **[1 mark]**

Figure 8

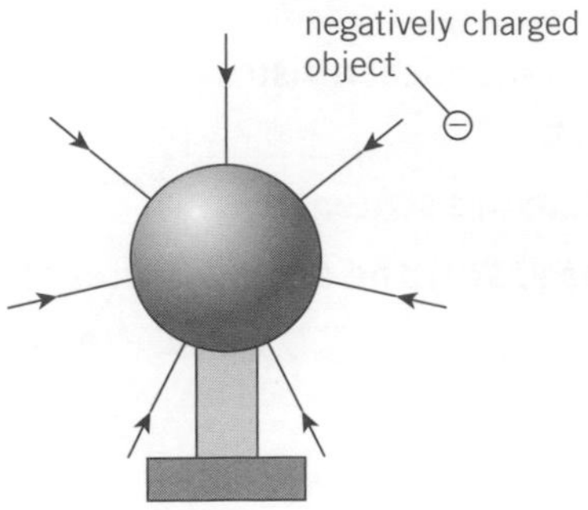

negatively charged object

07.3 Write down what happens to the force on the object as the student moves the object away from the sphere. **[1 mark]**

07.4 The student attaches a wire to the sphere.

They then attach the other end of the wire to a tap, which is connected to the ground.

A charge of 4×10^{-6} coulombs flows through the wire in a time of 0.2 s.

Calculate the average current in the wire. Show your working. **[4 marks]**

Exam Tip

If you're unsure on this question, look back at the definition of current in the Knowledge section.

08 Student A rubs a balloon on his jumper.

The balloon becomes negatively charged.

08.1 Name the particle that moves when objects become charged. **[1 mark]**

Exam Tip

Only the smallest subatomic particle moves!

08.2 Explain why the balloon becomes negatively charged. **[2 marks]**

08.3 Student A places the charged balloon on the wall.

It appears to 'stick' to the wall.

Student B says: 'the wall must be positively charged to make the balloon stick to it'.

Do you agree? Explain your answer. **[2 marks]**

Exam Tip

'Describe' in this question wants you to say *what* will happen, and 'Explain' wants you to say *why* it happened.

08.4 Student B rubs another balloon on the same jumper, and then holds both balloons up close together.

Describe what will happen to the balloons. Explain your answer. **[3 marks]**

09 A student has been given some samples of dough.

The dough conducts electricity because it has been made with salt.

The student connects one sample of dough in a series circuit with an ammeter and a battery marked 6 V, as shown in **Figure 9**.

The student then measures the current through the dough. The current is 15 mA.

Figure 9

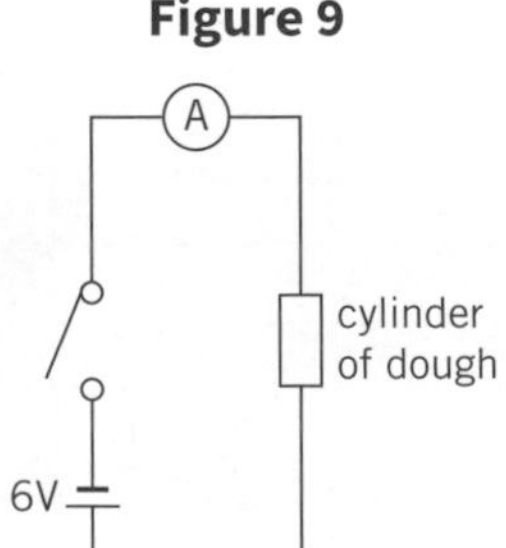

Exam Tip

Look out for the non-standard units.

09.1 Calculate the time it takes 0.6 C of charge to flow through the dough **[3 marks]**

09.2 The other samples of dough contain different masses of salt.

The student measures the current through these samples.

The data are shown in **Table 1**.

Exam Tip

Draw the graph in mA.

Trying to convert to amps will make the graph really small and hard to draw.

Table 1

Mass of salt per 100 g of dough in g	Current in mA
25	15
30	25
40	32
55	37
75	40
80	41

Sketch a graph of current against mass of salt from the data in **Table 1**. **[2 marks]**

Exam Tip

The only way to get full marks in the question is to use data from the table.

09.3 Use the data in **Table 1** to suggest how the resistance of the dough changes with the mass of salt.

Justify your answer with calculations. **[6 marks]**

Exam Tip

Don't worry if you've never come across lux before, it's just another unit!

10 The resistance of a light dependent resistor (LDR) changes with light intensity.

Light intensity is measured in lux.

10.1 Sketch a graph of the resistance of an LDR against light intensity. **[3 marks]**

Exam Tip

Sketch means you don't have to have numbers on your axes, but you do need to have labels!

10.2 Suggest a situation where you might need to use a light dependent resistor. **[1 mark]**

10.3 Using your answer to **10.2**, describe how the LDR can be used in that situation and explain why the changing resistance of the LDR would be useful.

You should include a circuit diagram to illustrate your answer. **[5 marks]**

Exam Tip

Your answer must refer to the situation you suggested in **10.2**!

11 Two students investigated the effect of length on the resistance of a wire.

They measured the resistance of different lengths of metal wire.

Table 2

Length in cm	Resistance in Ω
5	1.5
10	3.8
15	4.6
20	5.9
25	7.8

11.1 Plot the data from the table on graph paper. **[4 marks]**

Make sure you fill the space given as a tiny graph hidden in the corner is hard to read. If you make your work hard to read it is harder for the examiner to give you marks.

11.2 Identify the independent variable, the dependent variable, and **one** control variable. **[3 marks]**

11.3 Estimate the resistance of a piece of wire that is 22 cm long. **[1 mark]**

Exam Tip

Draw lines on your graph to show your working!

11.4 Suggest **one** improvement that the student could make to improve the precision of the data. **[1 mark]**

11.5 Student A looked at the graph and said 'the resistance is directly proportional to length'.

Student B looked at the graph and said 'there is a linear relationship between resistance and length'.

Which student has made a correct statement?

student A **student B** **both students**

Explain your answer. **[3 marks]**

Exam Tip

Use evidence from the graph in your answer.

12 A juggler is throwing a ball upwards and catching it.

12.1 Describe all the changes in the way that energy is stored from themoment the ball leaves the juggler's right hand, to the momentthe ball lands in his left hand after travelling through the air. **[4 marks]**

12.2 Write down the equation for calculating gravitational potential energy. **[1 mark]**

You will be given the value for gravity in the exam, but you still need to be able to remember the units.

12.3 The maximum change in gravitational potential energy is 0.4 J.

Gravitational field strength, g = 9.8 N/kg.

The mass of the ball is 0.1 kg.

Calculate the maximum height that the ball reaches. **[3 marks]**

13 **Table 3** shows data about two types of electric vehicle.

Table 3

Vehicle	Journey length in km	Useful energy transferred in kJ	Energy content of fuel used in kJ
A	100	4000	10 000
B	100	6000	

13.1 Calculate the efficiency of Vehicle **A**. Give your answer as a percentage. **[3 marks]**

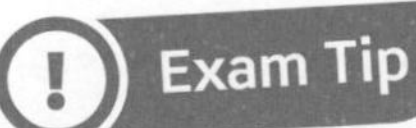

You may be used to giving the answer as a decimal. Remember to times that by 100 to get a percentage, and to add the % sign afterwards.

13.2 Vehicle **B** is twice as efficient as Vehicle **A**.

Calculate the energy content of the fuel used by Vehicle **B**. **[3 marks]**

13.3 Vehicle **B** completes the journey in a time of 2 hours.

Calculate the power of Vehicle **B**.

Give your answer to **two** significant figures. **[4 marks]**

14 A student sketches graphs of current against potential difference for three circuit components, **A**, **B**, and **C**, as shown in **Figure 10**.

> **Exam Tip**
> These graphs are a MUST LEARN!

Figure 10

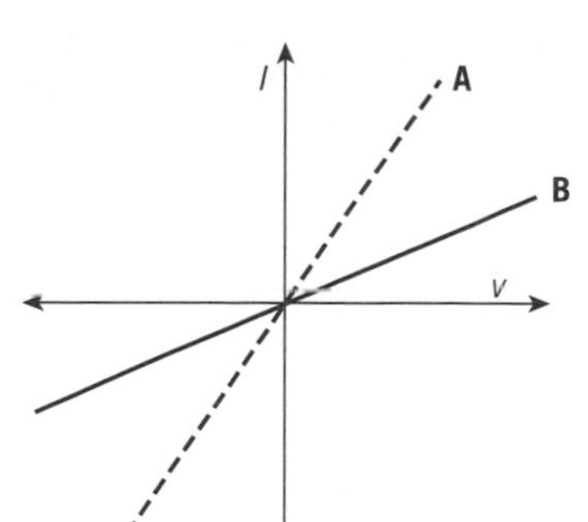

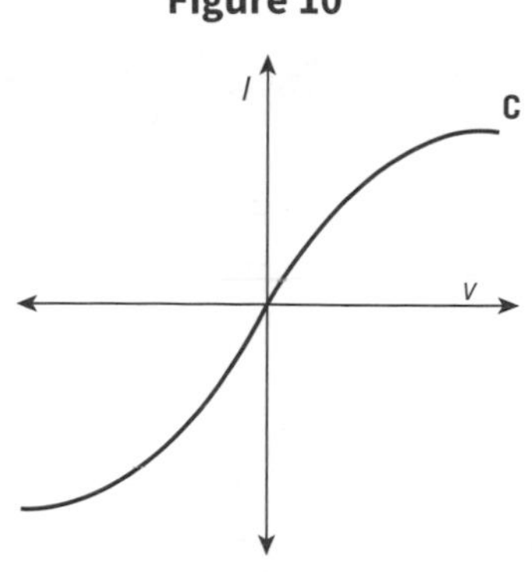

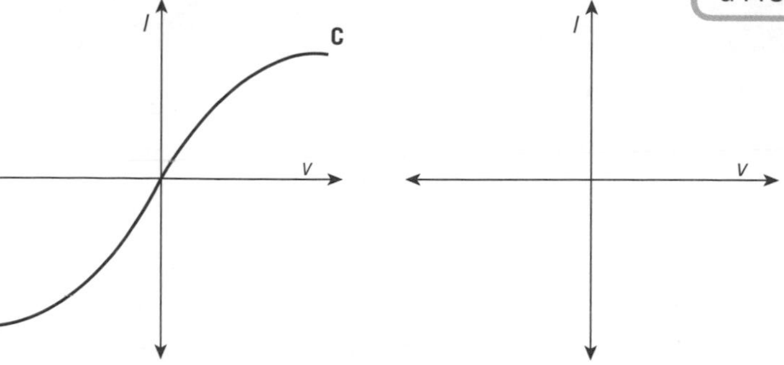

14.1 Match the descriptions with the correct graph from **Figure 10**. **[2 marks]**

Description	Graph
the graph for a resistor with a large resistance	**A**
the graph for a filament lamp	**B**
the graph for a resistor with a small resistance	**C**

14.2 On the empty axes in **Figure 10**, sketch the graph of current against potential difference for a diode. **[2 marks]**

14.3 A student connects some LEDs in a circuit as shown in **Figure 11**.

Write down which LEDs (**X**, **Y**, **Z**), if any, will light up.

Explain your answer. **[2 marks]**

Figure 11

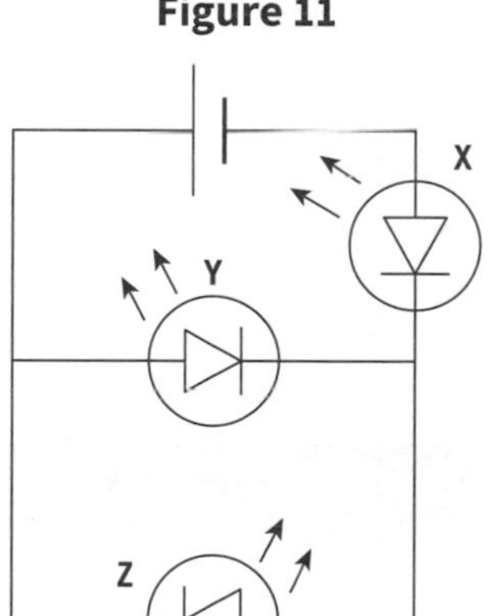

> **Exam Tip**
> To answer this question, think about what the symbol for a diode is actually showing.

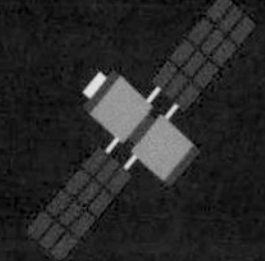

Knowledge

P6 Energy of matter

Changes of state and states of matter

Changes of state and conservation of mass

Changes of state are physical changes because no new substances are produced. The mass always stays the same because the number of particles does not change.

Particles and kinetic energy

When the temperature of a substance is increased, the kinetic energy store of its particles increases and the particles vibrate or move faster.

If the kinetic store of a substance's particles increases or decreases enough, the substance may change state.

Density

You can calculate the density of an object if you know its mass and volume:

$$\text{density (kg/m}^3\text{)} = \frac{\text{mass (kg)}}{\text{volume (m}^3\text{)}}$$

$$\rho = \frac{m}{V}$$

Gas	**Arrangement**	• particles are spread out • almost no forces of attraction between particles • large distance between particles on average
	Movement	• particles move randomly at high speed
	Properties	• low density • no fixed volume or shape • can be compressed and can flow • spread out to fill all available space
Liquid	**Arrangement**	• particles are in contact with each other • forces of attraction between particles are weaker than in solids
	Movement	• particles are free to move randomly around each other
	Properties	• usually lower density than solids • fixed volume • shape is not fixed so they can flow
Solid	**Arrangement**	• particles held next to each other in fixed positions by strong forces of attraction
	Movement	• particles vibrate about fixed positions
	Properties	• high density • fixed volume • fixed shape (unless deformed by an external force)

Internal energy

Heating a substance increases its **internal energy**.

Internal energy is the sum of the total kinetic energy the particles have due to their motion and the total potential energy the particles have due to their positions relative to each other.

Latent heat

In a graph showing the change in temperature of a substance being heated or cooled, the flat horizontal sections show when the substance is changing state.

The energy transfers taking place during a change in state do not cause a change in temperature, but do change the internal energy of the substance.

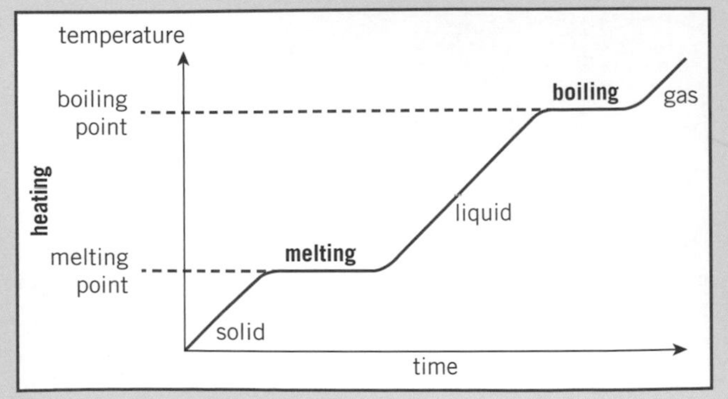

The energy transferred when a substance changes state is called the **latent heat**.

Specific latent heat – the energy required to change 1 kg of a substance with no change in temperature.

Specific latent heat of fusion – the energy required to melt 1 kg of a substance with no change in temperature.

Specific latent heat of vaporisation – the energy required to evaporate 1 kg of a substance with no change in temperature.

The energy needed to change the state of a substance can be calculated using the equation:

$$\text{thermal energy for a change in state (J)} = \text{mass (kg)} \times \text{specific latent heat (J/kg)}$$

$$E = m \times l$$

The relationship between temperature and pressure in gases

Gas temperature

The particles in a gas are constantly moving in random directions and with random speeds.

The temperature of a gas is related to the average kinetic energy of its particles.

When a gas is heated, the particles gain kinetic energy and move faster, so the temperature of the gas increases.

Gas pressure

The pressure a gas exerts on a surface, such as the walls of a container, is caused by the force of the gas particles hitting the surface.

The pressure of a gas produces a net force at right angles to the walls of a container or any surface.

If the temperature of a gas in a sealed container is increased, the pressure increases because

- the particles move faster so they hit the surfaces with more force
- the number of these impacts per second increases, exerting more force overall.

If a gas is compressed quickly, for example, in a bicycle pump, its temperature can rise. This is because

- compressing the gas requires a force to be applied to the gas – this results in work being done to the gas, since work done = force × distance
- the energy gained by the gas is not transferred quickly enough to its surroundings.

The relationship between volume and pressure in gases

If the volume of a fixed mass of gas at a constant temperature is decreased, the pressure increases because

- the distance the particles travel between each impact with a container wall is smaller
- the number of impacts per second increases, so the total force of impacts increases.

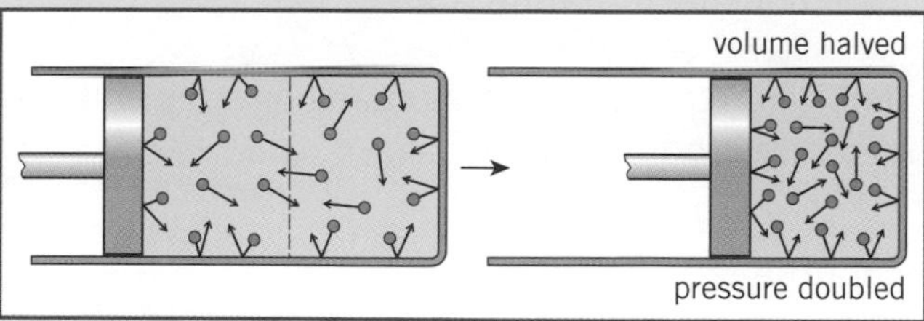

Similarly, if the volume is increased, the pressure decreases. This is because

- the distance the particles travel between each impact with a wall of the container is greater
- the number of impacts per second decreases, so the total force of the impacts decreases.

The pressure and volume of a fixed mass of gas at a constant temperature are linked by the equation:

$$\text{pressure (Pa)} \times \text{volume (m}^3\text{)} = \text{constant}$$

$$p \times V = \text{constant}$$

Rearranging this equation gives:

$$p = \frac{\text{constant}}{V} \quad \text{and} \quad V = \frac{\text{constant}}{p}$$

This shows that pressure is inversely proportional the volume of a gas.

Revision tips

Practice Draw diagrams to show the arrangement of particles in solids, liquids, and gases, with labels to show the movement and forces of attraction. Cover up the page opposite and write down the properties of each type of substance.

Remember In the equation $p = V \times \text{constant}$, the constant is just a number so doesn't have any units.

Remember Pressure (p) and density (ρ) have very similar looking symbols. ρ is the lower case of the Greek letter rho. Don't get them confused!

Key terms

Write a definition for these key terms.

boiling condensation conservation of mass density evaporation freezing fusion internal energy latent heat melting specific latent heat sublimation vaporisation

Retrieval

Learn the answers to the questions below then cover the answers column with a piece of paper and write as many as you can. Check and repeat.

	P6 questions	Answers
1	Which two quantities do you need to measure to find the density of a solid or liquid?	mass and volume
2	What happens to the particles in a substance if its temperature is increased?	they move faster and the energy in their kinetic energy store increases
3	Why are changes of state physical changes?	no new substances are produced and the substance will have the same properties as before if the change is reversed
4	Why is the mass of a substance conserved when it changes state?	the number of particles does not change
5	What is the internal energy of a substance?	the total kinetic energy and potential energy of all the particles in the substance
6	Why does a graph showing the change in temperature as a substance cools have a flat section when the substance is changing state?	the energy transferred during a change in state causes a change in the internal energy of the substance
7	What is the name given to the energy transferred when a substance changes state?	latent heat
8	What is the specific latent heat of a substance?	the energy required to change the state of one kilogram of that substance with no change in temperature
9	What is the specific latent heat of fusion a substance?	the energy required to change one kilogram of the substance from solid to liquid at its melting point, without changing its temperature
10	What is the specific latent heat of vaporisation of a substance?	the energy required to change one kilogram of the substance from liquid to vapour at its boiling point, without changing its temperature
11	On a graph of temperature against time for a substance being heated up or cooled down, what do the flat (horizontal) sections show?	the time when the substance is changing state and the temperature is not changing
12	What property of a gas is related to the average kinetic energy of its particles?	temperature
13	What causes the pressure of a gas on a surface?	the force of the gas particles hitting the surface
14	Give two reasons why the pressure of a gas in a sealed container increases if its temperature is increased.	the molecules move faster so they hit the surfaces with more force and the number of impacts per second increases, so the total force of the impacts increases
15	Give two reasons why the temperature of a gas increases if it is compressed quickly.	the force applied to compress the gas results in work being done to the gas, and the energy gained by the gas is not transferred quickly enough to the surroundings
16	Explain why the pressure of a fixed mass of gas decreases if the volume is increased and kept at constant temperature.	the distance the particles travel between each impact with a wall of the container is greater, so the number of impacts per second decreases, so the total force of the impacts decreases

Now go back and use the questions below to check your knowledge from previous chapters.

Previous questions

Answers

	Previous questions	Answers
1	What is the same at all points when charge flows in a closed loop?	current
2	What are the units for power, current, potential difference, and resistance?	watts (W), amps (A), volts (V), ohms (Ω)
3	What colours are the live, neutral, and earth wires in a three-core cable?	live = brown, neutral = blue, earth = green and yellow stripes
4	What is an ohmic conductor?	conductor where current is directly proportional to the voltage so resistance is constant (at constant temperature)
5	What are the main disadvantages of using tidal power?	can harm marine habitats, initial set up expensive, cannot increase supply when needed, amount of energy varies on time of month, hazard for boats
6	What factors affect the rate of heat loss from a building?	thickness of walls and roof, thermal conductivity of walls and roof, the temperature difference between the two sides of the wall/roof

Put paper here

Put paper here

Put paper here

Required Practical

Practise answering questions on the required practicals using the example below. You need to be able to apply your skills and knowledge to other practicals too.

Investigate the density of regular and irregular shaped solids and liquids

Density

You need to be able to measure the masses and volumes of regularly and irregularly-shaped solid objects, and liquids.

To be accurate and precise in your investigation you need to:

- use dimensions to determine volume of regularly-shaped objects, and displacement for irregularly-shaped objects
- use a measuring cylinder to measure the volume of a liquid or displaced water
- choose a measuring cylinder that is just large enough for the object/liquid so that the divisions are as small as possible
- measure from the bottom of the meniscus of a liquid in a measuring cylinder.

Worked example

A student uses a measuring cylinder and digital balance to measure the density of an irregular lump of putty.

1 Describe how to use the equipment to make the measurements required.

Place the putty in the cylinder and cover it with water so that the putty is just covered. Record the volume, remove the putty, and record the new volume. Find the difference between the volumes in cm^3. Measure the putty mass with the digital balance in g.

2 For a putty mass of 65 g, the student recorded three volumes of 54, 56, and 57 cm^3. Calculate the mean volume and the density of the putty.

$$\text{mean volume} = \frac{(54 + 56 + 57)}{3} = 55.67\ \text{cm}^3$$

$$\text{density} = \frac{\text{mass}}{\text{volume}} = \frac{65}{55.67} = 1.17\ \text{g/cm}^3$$ to 3 significant figures

Practice

A scientist has two samples of seawater. The volume of each is the same.

Sample **A** has a mass of 235 g and Sample **B** has a mass of 237 g.

Calculate the ratio of the densities of Sample **A** and **B**.

Show your working and explain your method.

Practice

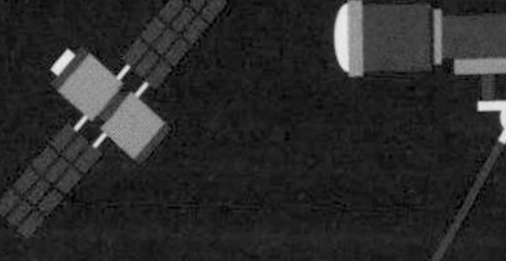

Exam-style questions

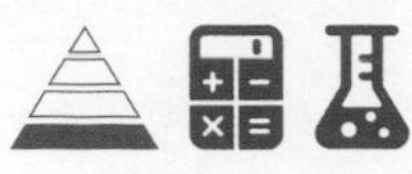

01 A student wants to calculate the density of modelling clay.

To do this, they take a mass of clay and put it into a measuring cylinder containing water.

Figure 1 shows the water in the measuring cylinder before (**A**) and after (**B**) the clay was added.

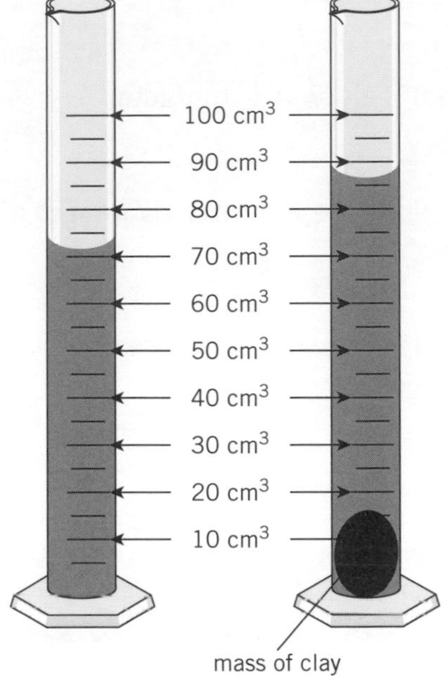

Figure 1

01.1 Use **Figure 1** to calculate the volume of the clay. **[2 marks]**

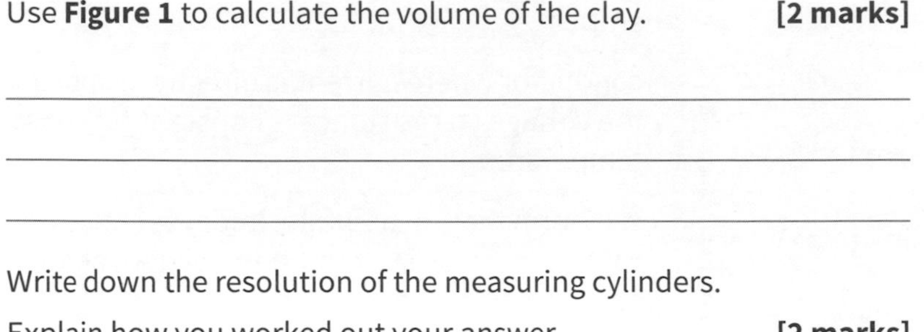

01.2 Write down the resolution of the measuring cylinders.

Explain how you worked out your answer. **[2 marks]**

01.3 The student measures the mass of the clay.

The clay has a mass of 23.41 g.

Suggest the measuring instrument that the student used to find the mass. **[1 mark]**

______ cm^3

01.4 Write down the equation that links density, mass, and volume. **[1 mark]**

01.5 Calculate the density of the clay in g/cm^3. **[2 marks]**

______ g/cm^3

01.6 Another student makes a cube out of the clay.

Suggest a method that this student could use to find the volume of the clay. **[2 marks]**

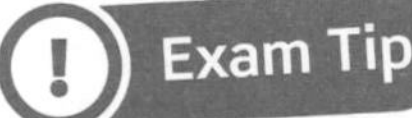

Exam Tip

An easy way to remember the equation in **01.4** is that the m and v can look like a heart:

Exam Tip

A cube is a regular shape – you don't need to use a measuring cylinder for this.

02 A student is learning about internal energy.

They draw two diagrams, **A** and **B**, as shown in **Figure 2**.

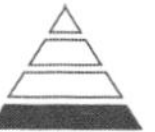

02.1 Complete the sentences using the words in the box. **[4 marks]**

kinetic	vibrating	moving fast
potential	gravitational	moving slowly

In diagram **A** the particles are ________________. Most of the internal energy is due to the ________________ energy of the particles.

In diagram **B** the particles are ________________. Most of the internal energy is due to the ________________ energy of the particles.

Figure 2

A

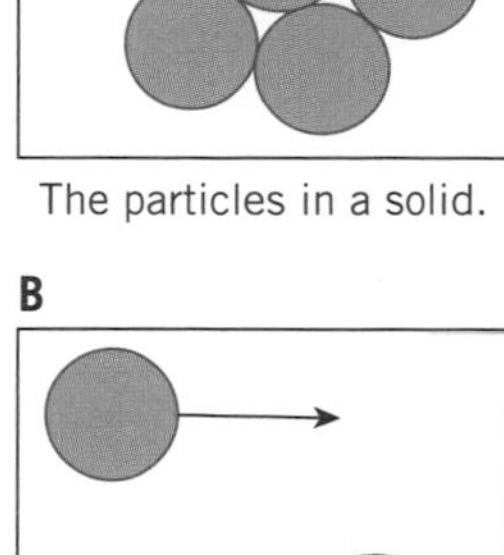

The particles in a solid.

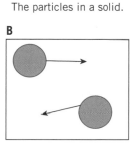

B

The particles in a gas.

02.2 The sample shown in **Figure 2 A** is heated for a long time.

Describe how the internal energy of the sample changes. **[2 marks]**

____

____

____

02.3 The sample shown in **Figure 2 B** is heated.

The student decides to use the particle model to describe and explain what happens.

Which statement is correct?

Tick **one** box. **[1 mark]**

As the gas is heated the average kinetic energy of the molecules decreases. ☐

The average kinetic energy of the molecules is independent of the temperature of the gas. ☐

If the temperature of a gas increases the pressure that the gas exerts decreases (if the volume stays the same). ☐

The particles in a gas are in random motion. ☐

Exam Tip

Only tick *one* box, if you tick more boxes you won't get any marks!

03 A liquid is heated until it boils. The temperature is measured as it is being heated.

Figure 3 shows the graph of temperature against time for the liquid.

Figure 3

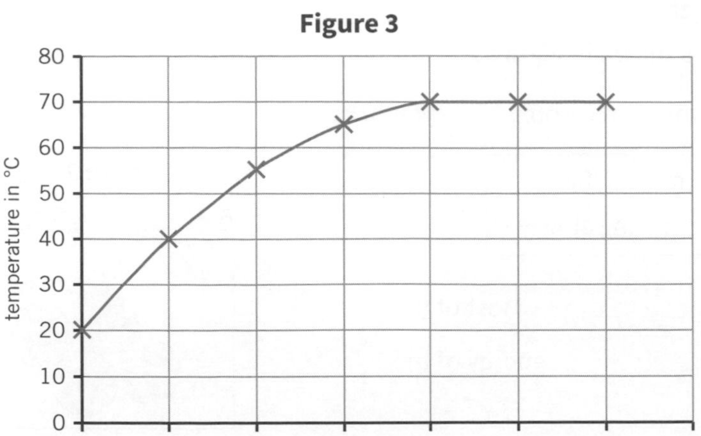

03.1 Write down the temperature of the room. **[1 mark]**

03.2 Write down the boiling point of the liquid. **[1 mark]**

03.3 Give the equation that links power, energy transferred, and time. **[1 mark]**

03.4 The power of the heater used to heat the liquid was 1 kW. The specific latent heat of vaporisation of the liquid is 365 kJ/kg. Calculate the mass of liquid that was vaporised between 4 and 6 minutes. Use an equation from the *Physics Equations Sheet*. **[5 marks]**

03.5 Suggest whether your answer to part **09.4** is an overestimate or an underestimate of the actual mass of liquid that evaporated. Explain your answer. **[2 marks]**

03.6 Sketch a line on **Figure 3** to show what would happen if the power of the heater was doubled. **[3 marks]**

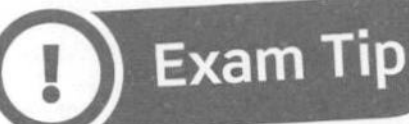

Exam Tip

In question **03.1** we can assume that the liquid was at room temperature before it started to be heated.

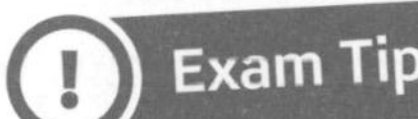

Exam Tip

You need to use more than one equation to solve question **03.4**.

04 A teacher is showing a class a method for finding the specific latent heat of vaporisation of water. The teacher puts a kettle containing water on a set of digital scales and measures its mass.

The kettle is turned on to allow the water to boil. At the same time, the teacher turns on a stopwatch. After two minutes the kettle is turned off and the teacher notes the new reading on the scales.

Table 1

mass at the start of the two minutes	1.276 kg
mass at the end of the two minutes	1.180 kg

The power of the kettle is 2 kW.

04.1 Write down the equation that links energy, power, and time. **[1 mark]**

04.2 Calculate the energy transferred from the kettle to the water. **[4 marks]**

04.3 Use your answer to **04.2** and the correct equation from the *Physics Equations Sheet* to calculate the specific latent heat of vaporisation of water.

Give your answer in kJ/kg. **[5 marks]**

> **Exam Tip**
> Look at the difference in values in **Table 1**.

04.4 The textbook value for the specific latent heat of vaporisation of water is 2265 kJ/kg. Suggest a reason for the difference between the value that you have calculated and the textbook value.

Explain your answer. **[3 marks]**

05 A student is looking at a helium gas tank. The label says 'Caution: gas under pressure'.

05.1 Explain what 'under pressure' means. **[1 mark]**

05.2 The student wants to investigate the behaviour of gases using a syringe. They hold the end of the syringe to seal it and then push the plunger down quickly. After doing this several times, the student notices that the gas gets hot.

Explain in terms of 'doing work' why the gas gets hot. **[2 marks]**

05.3 Explain in terms of particles why the gas gets hot. **[2 marks]**

6 A student noticed that when they finished having a shower, the mirror is 'fogged up'.

06.1 Explain in terms of energy why the mirror is covered by a thin layer of water. **[3 marks]**

> **Exam Tip**
> First use the equation for specific latent heat of vaporisation to find the mass of water.

06.2 The student estimates that the mirror is a square with sides measuring 60 cm. The density of water is 1×10^3 kg/m^3. The specific latent heat of vaporisation of water is 2265 kJ/kg. While the fog was forming, a total of 730 kJ of energy was transferred.

Calculate the thickness of the layer of water on the mirror. **[6 marks]**

07 A gas can be compressed or expanded by changes in pressure.

07.1 Describe how the arrangement of particles in a gas is different from that in a solid. **[4 marks]**

07.2 Explain, in terms of particles, why a gas exerts a pressure. **[4 marks]**

07.3 Sketch a graph to show how the pressure of a fixed mass of gas at constant volume varies with temperature. **[3 marks]**

08 One way to heat milk is to pass steam through it.

08.1 Suggest how a jet of steam heats a cup of milk. **[2 marks]**

08.2 The mass of milk in a cup is 242 g. The specific heat capacity of milk is 3.93 kJ/kg °C. Show that the energy required to heat the milk from 20 °C to 70 °C is about 48 kJ. Use an equation from the *Physics Equations Sheet*. **[4 marks]**

> **Exam Tip**
> Look out for non-standard units!

08.3 The specific latent heat of vaporisation of water is 2260 kJ/kg. Calculate the mass of steam that would need to condense into water to produce the energy calculated in **08.2** **[4 marks]**

08.4 Write down one assumption that you made when doing the calculation. **[1 mark]**

09 A substance is heated. **Figure 4** shows how the temperature of the substance changes with time. The straightline sections of the graph are labelled **A, B, C, D,** and **E**.

Figure 4

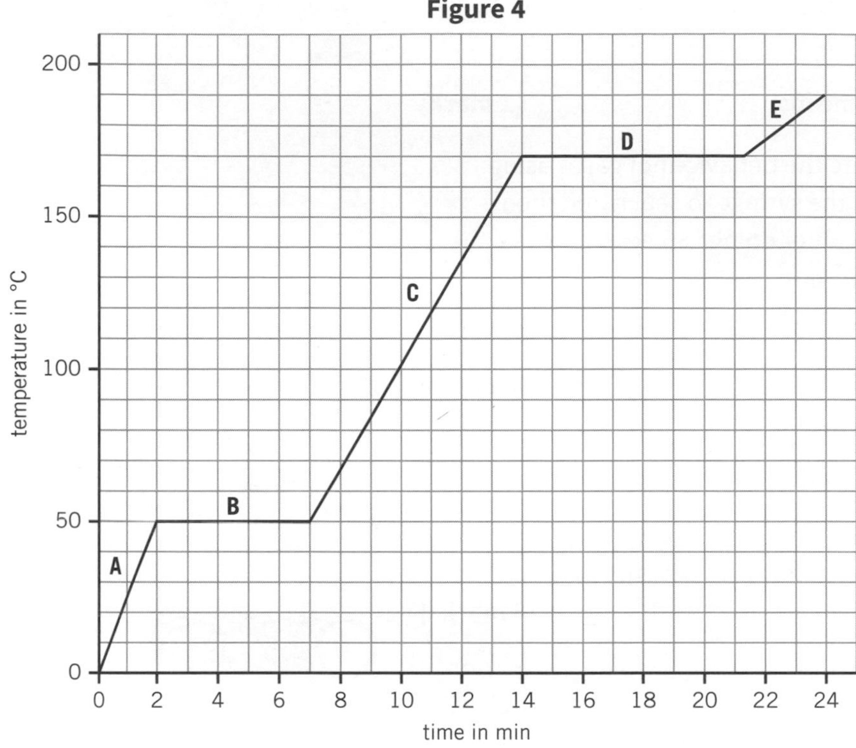

09.1 Write the letters of all the sections of the graph that show a change of state.

Explain why you have chosen these sections. **[4 marks]**

09.2 Did the substance start out as a solid or a liquid?

Explain your answer. **[2 marks]**

09.3 Write down the section of the graph where the vibration of the particles is increasing. **[1 mark]**

09.4 Write down the **two** sections of the graph where the kinetic energy of the particles is increasing. **[2 marks]**

Exam Tip

Question **09.1** is worth four marks – it gives you a clue to what the examiner is looking for and helps structure your answer:

- 1st mark, give letter showing change of state
- 2nd mark, explain why you chose that letter
- 3rd mark, give letter showing second change of state
- 4th mark, explain why you have chosen this letter

10 A teacher has a cube of dry ice. Dry ice is solid carbon dioxide. The length of the sides of the cube is 5 cm.

10.1 Calculate the volume of the cube in cm^3. **[2 marks]**

10.2 The density of dry ice is 1.5 g/cm^3. Calculate the mass of the cube of dry ice. **[4 marks]**

The units in question **10.2** give you a big clue to the equation you need!

10.3 The dry ice changes from a solid to a gas without going through a liquid phase. Write down the name of this change of state. **[1 mark]**

10.4 Explain why the change is a physical change and not a chemical change. **[1 mark]**

10.5 Compare the internal energy of the carbon dioxide gas with the internal energy of the dry ice.

Explain your answer. Assume the gas and solid are at the same temperature. **[2 marks]**

11 A student wants to find out what happens when you put the human body into very cold water. To model what happens when the human body is put into cold water, the student uses a test tube filled with water and places it into a large beaker of iced water.

The student sets up a circuit with a sensor to monitor the temperature of the water in the test tube over a period of time. The sensor is placed in the water in the test tube.

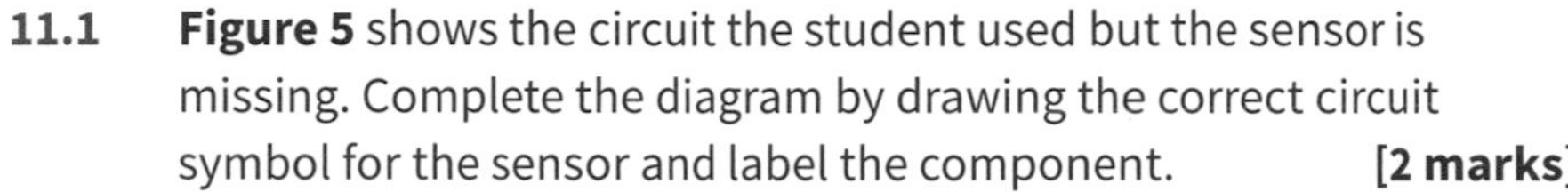

11.1 **Figure 5** shows the circuit the student used but the sensor is missing. Complete the diagram by drawing the correct circuit symbol for the sensor and label the component. **[2 marks]**

Figure 5

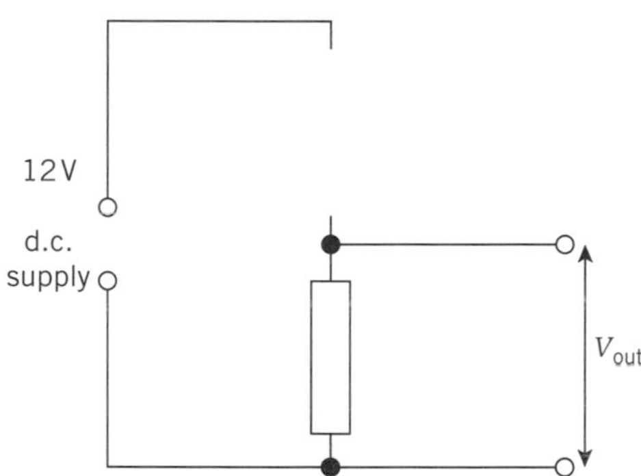

11.2 The student connects up the circuit. Initially $V_{out} = 6\,V$. Describe what will happen to V_{out} when they put the test tube into a large beaker of iced water. Explain your answer. **[5 marks]**

11.3 A data logger is used to monitor V_{out}. Eventually the data logger shows that V_{out} is no longer changing. Describe how the student can use this measurement to calculate the temperature of the water. Include any additional data the student needs in order to do the calculation. **[4 marks]**

Exam Tip

You don't need to do the actual calculation, just describe how you would do it and what numbers you would need.

11.4 Suggest **one** limitation of this model for working out what happens to the temperature of a human body in cold water. **[1 mark]**

12 A light bulb is powered by a generator. The power of the light bulb is 0.24 W. The generator is powered by a falling 300 g mass attached to the generator by a string, as shown in **Figure 6**.

Figure 6

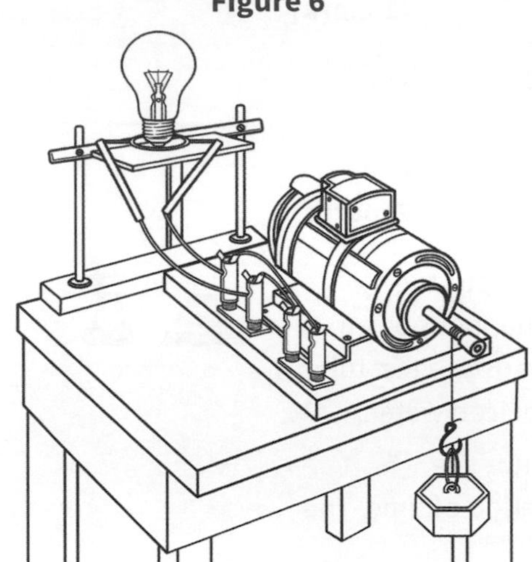

The generator is 90 % efficient. The bulb needs to be powered at full brightness for 1 minute. Calculate how far the mass must fall.

Explain the reasons for your calculations. **[6 marks]**

13 A student uses two potatoes to make a battery. The battery is then used to run a small clock. The student measures the potential difference produced by the battery.

13.1 Name the measuring instrument that is used to measure potential difference. **[1 mark]**

Exam Tip

In **13.1** think about the units for potential difference.

13.2 Write down the equation that links potential difference, current, and resistance. **[1 mark]**

13.3 Calculate the potential difference that you need to produce a current of 0.15 A in a clock with a resistance of 10 Ω. **[2 marks]**

13.4 Write down the equation that links charge flow, current, and time. **[1 mark]**

13.5 Calculate the charge that flows through the clock every minute. Write the correct unit with your answer. **[3 marks]**

Exam Tip

The standard unit for time is seconds, so remember to convert minutes to seconds before you start question **13.5**.

13.6 A student decides to use four potatoes instead of one. Describe what would happen to the current flowing in the clock.

Give a reason for your answer. **[2 marks]**

14 A teacher demonstrates what happens when you remove the air from a container containing marshmallows. **Figure 7** shows the apparatus used.

Figure 7

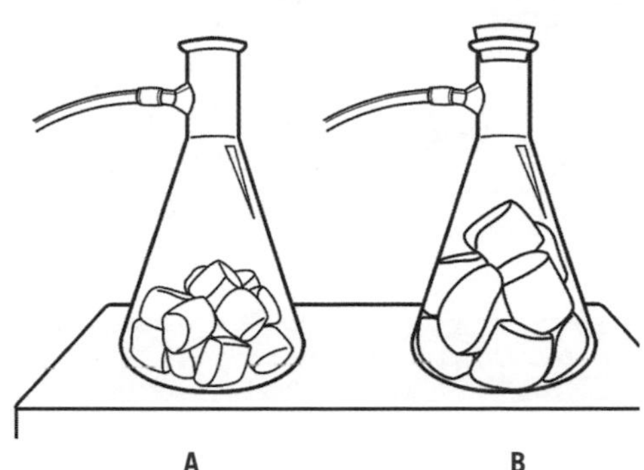

14.1 Marshmallows are full of tiny pockets of air. Explain why the marshmallows increase in size.

Use ideas about forces, pressure, and the particle model in your answer. **[5 marks]**

14.2 The marshmallows in **Figure 7 A** are at room temperature and pressure (101 kPa). The volume of a marshmallow is 8×10^{-6} m^3. The volume of a marshmallow in **Figure 7 B** is 2.7×10^{-5} m^3.

Calculate the change in pressure of a marshmallow when the teacher pumps out the air. **[6 marks]**

Exam Tip

Standard form comes up frequently – it is worth spending the time to become familiar with writing it and putting it into your calculators.

Knowledge

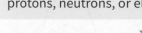

P7 Atoms

Modern model of an atom

The model of the atom we have today was developed over time with the help of evidence from experiments.

Future experiments may change our understanding and lead us to alter or replace this model of the atom.

Dalton's model

John Dalton thought of the atom as a solid sphere that could not be divided into smaller parts. His model did not include protons, neutrons, or electrons.

Plum pudding model

Scientists' experiments resulted in the discovery of charged sub-atomic particles. The first to be discovered were electrons – tiny, negatively charged particles.

The discovery of electrons led to the **plum pudding model** of the atom – a cloud of positive charge, with negative electrons embedded in it.

Protons and neutrons had not yet been discovered.

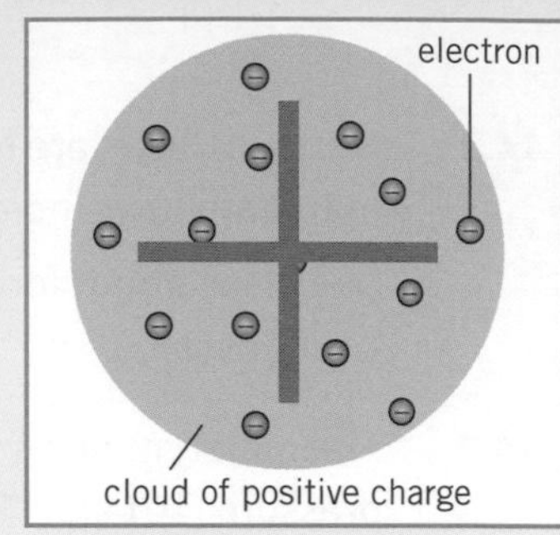

Alpha scattering experiment

1. Scientists fired small, positively charged particles (called **alpha particles**) at a piece of gold foil only a few atoms thick.
2. They expected the alpha particles to travel straight through the gold.
3. They were surprised that a few of the alpha particles bounced back and some were deflected (alpha scattering).
4. To explain why the alpha particles were repelled, the scientists suggested that the positive charge and mass of an atom must be concentrated in a small space at its centre. They called this space the nucleus.

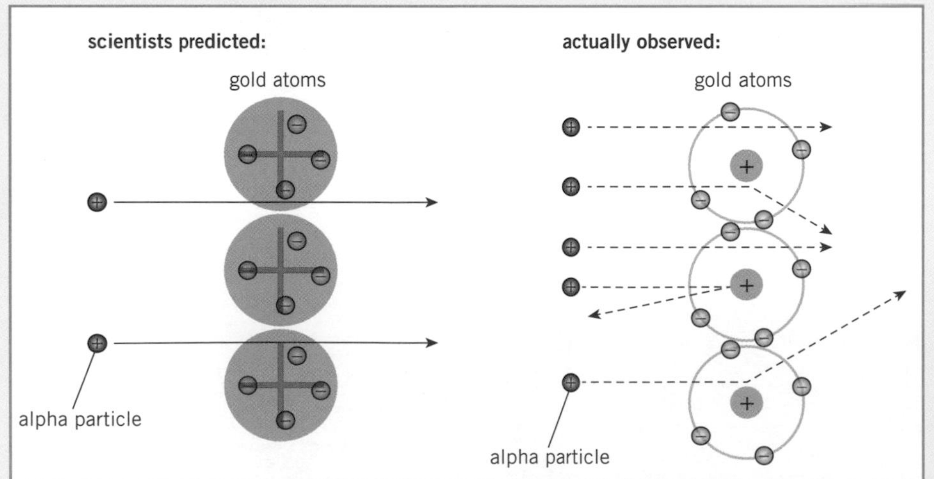

Nuclear model

Scientists replaced the plum pudding model with the **nuclear model**. They suggested that the electrons **orbit** (go around) the nucleus, but not at set distances, and the mass of the atom was concentrated in the charged nucleus.

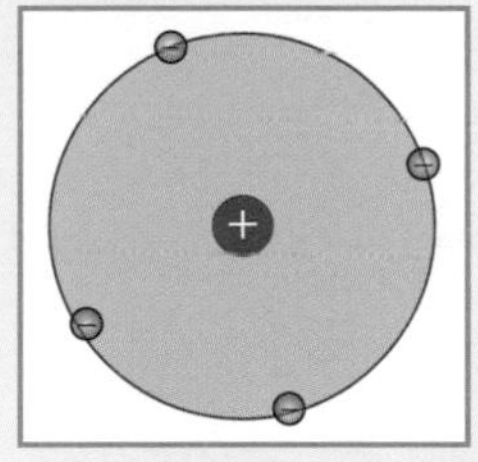

Bohr's model

Niels Bohr improved the nuclear model, and calculated that electrons must orbit the nucleus at fixed distances. These orbits are called shells or energy levels. These calculations agreed with experimental results.

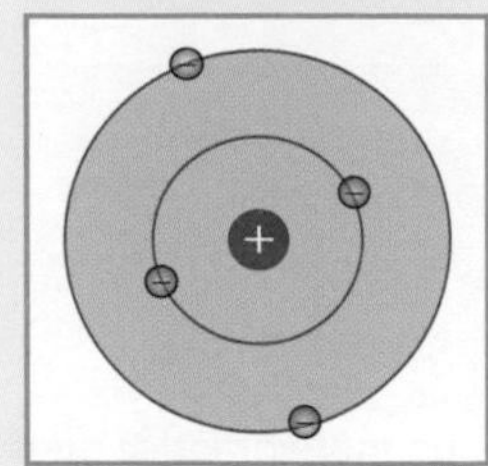

Revision tip

This content is also in chemistry. This can be a good thing and a bad thing – good because you have to learn less, and bad because it can be confusing in exams.

A few years ago there was a question about the structure of an atom in physics. If it had been in a chemistry exam everyone would have done brilliantly, but students panicked.

Protons

Later experiments provided evidence that the positive charge of a nucleus could be split into smaller particles, each with an opposite charge to the electron. These positively charged particles were called protons.

Neutrons

James Chadwick carried out experiments that provided evidence for a particle with no charge. Scientists called this the neutron. They concluded that the protons and neutrons are in the nucleus, and the electrons orbit the nucleus in shells.

Basic structure of an atom

An **atom** has a radius of about 1×10^{-10} metres.

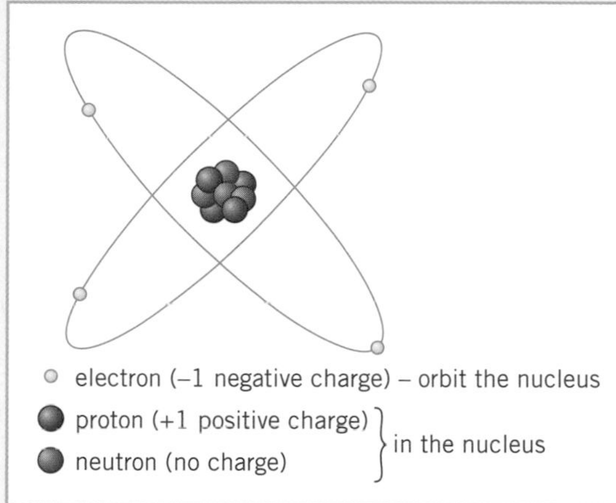

electron (–1 negative charge) – orbit the nucleus
proton (+1 positive charge)
neutron (no charge)
in the nucleus

An atom is uncharged overall, and has equal numbers of **protons** and **electrons**.

The nucleus

- Has a radius about 10 000 times smaller than the radius of an atom
- Contains protons and **neutrons**
- Is where most of the mass of an atom is concentrated.

Electrons

- Orbit the nucleus at different fixed distances called **energy levels**.
- Can gain energy by absorbing electromagnetic radiation. This causes them to move to a higher energy level.
- Can lose energy by emitting electromagnetic radiation. This causes them to move to a lower energy level.

Element symbols

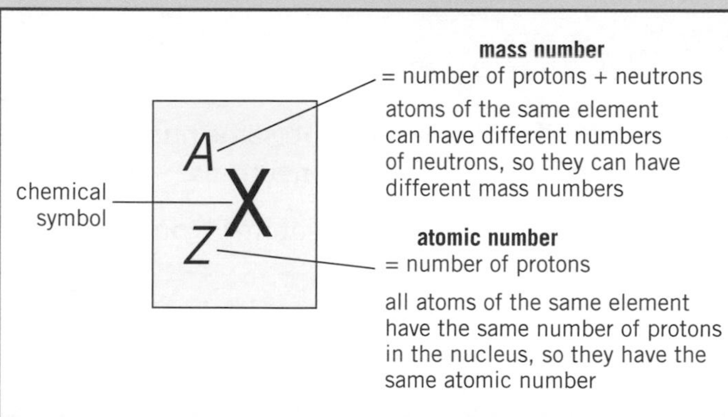

mass number
= number of protons + neutrons
atoms of the same element can have different numbers of neutrons, so they can have different mass numbers
chemical symbol
A
X
Z
atomic number
= number of protons
all atoms of the same element have the same number of protons in the nucleus, so they have the same atomic number

Isotopes are atoms of the same element, with the same number of protons but a different numbers of neutrons.

Electrons

Atoms can become charged when they lose or gain electrons. This process is called **ionisation**.

- A positive ion is formed if an uncharged atom loses one or more electrons.
- A negative ion is formed if an uncharged atom gains one or more electrons.

Key terms

Make sure you can write a definition for these key terms.

alpha particle, atom, atomic number, electron, energy level, ionisation, isotope, mass number, neutron, orbit, plum pudding model, proton

Retrieval

Learn the answers to the questions below then cover the answers column with a piece of paper and write as many as you can. Check and repeat.

	P7 questions	Answers
1	Describe the basic structure of an atom.	nucleus containing protons and neutrons, around which electrons orbit in fixed energy levels/shells
2	Describe the plum pudding model of the atom.	sphere of positive charge with negative electrons embedded in it
3	What charges do protons, neutrons, and electrons carry?	protons = positive, neutrons = no charge, electrons = negative
4	Why do atoms have no overall charge?	equal numbers of positive protons and negative electrons
5	What is the radius of an atom?	around 1×10^{-10} m
6	How small is a nucleus compared to a whole atom?	around 10 000 times smaller
7	How can an electron move up an energy level?	absorb sufficient electromagnetic radiation
8	What is ionisation?	process which adds or removes electrons from an atom
9	What is formed if an atom loses an electron?	positive ion
10	How does an atom become a negative ion?	gains one or more electrons
11	What is the atomic number of an element?	number of protons in one atom of the element
12	What is the mass number of an element?	number of protons + number of neutrons
13	Which particle do atoms of the same element always have the same number of?	protons
14	What are isotopes?	atoms of the same element (same number of protons) with different numbers of neutrons
15	What were the two main conclusions from the alpha particle scattering experiment?	• most of the mass of an atom is concentrated in the nucleus • nucleus is positively charged

Put paper here

Now go back and use the questions below to check your knowledge from previous chapters.

Previous questions | Answers

	Previous questions	Answers
1	What is the relationship between the temperature of an object and its emission of infrared radiation?	the higher the temperature of an object, the more infrared radiation emitted in a given time
2	Why are changes of state physical changes?	no new substances are produced and the substance will have the same properties as before if the change is reversed
3	What is the function of the earth wire in a three-core cable?	safety wire to stop the appliance becoming live
4	What are the main features of a parallel circuit?	p.d. across each branch is the same, total current through circuit is the sum of the currents in each branch – total resistance of all resistors is less than the resistance of the smallest individual resistor
5	How does a material become charged?	becomes negatively charged by gaining electrons and becomes positively charged by losing electrons
6	How does having a large potential difference in the transmission cables help to make the National Grid an efficient way to transfer energy?	large p.d. means a small current is needed to transfer the same amount of power, small current in the transmission cables means less electrical power is wasted due to heating

Put paper here

Maths Skills

Practise your maths skills using the worked example and practice questions below.

Order of magnitude

Orders of magnitude are useful for comparing the size of numbers.

An order of magnitude is a factor of 10, so it is usually written as 10^n.

For example, if one number is roughly 10 times bigger than another number, it is one order of magnitude bigger.

If a number is 1000 times bigger than another number, it is three orders of magnitude bigger, because: $1000 = 10 \times 10 \times 10 = 10^3$.

Two numbers are of the same order of magnitude if dividing the bigger number by the smaller number gives an answer less than 10. For example, 12 and 45 are the same order of magnitude, but 13 and 670 are not.

If numbers are written in standard form, their orders of magnitude can be compared by dividing the larger power of ten by the smaller power of ten.

Worked Examples

A mouse has a mass of about 40 g, and an elephant has a mass of about 4×10^6 g. How many orders of magnitude heavier is an elephant compared to the mouse?

Answer: Divide the larger power of ten by the smaller power of ten:

$\frac{10^6}{10^1} = 10^5$, or five orders of magnitude.

The mass of the Sun is about 2.0×10^{30} kg. The mass of the Earth is 6.0×10^{24} kg. How many orders of magnitude bigger is the Sun's mass?

Answer: Divide the larger power of ten by the smaller power of ten:

$\frac{10^{30}}{10^{24}} = 10^6$, or six orders of magnitude.

Practice

1 How many orders of magnitude bigger is the mass of a lorry around 44 000 kg compared to a human body around 70 kg?

2 How many orders of magnitude smaller is the diameter of an atom around 0.1 nm compared to the diameter of a marble of 1 cm?

3 How many orders of magnitude smaller is 70 J compared to 450 MJ?

Practice

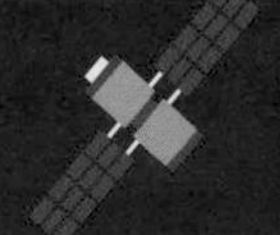

Exam-style questions

01 **Table 1** shows the number of protons and neutrons in the neutral atoms of three elements.

Table 1

Element	Number of protons	Number of neutrons
A	10	10
B	10	12
C	11	12

01.1 Write down the number of electrons in an atom of element **A**. **[1 mark]**

01.2 Explain your answer to **01.1**. **[2 marks]**

Use data from the table.

01.3 Name the two elements that are isotopes. **[1 mark]**

Isotopes have the same atomic number.

01.4 Explain your answer to **01.3**. **[2 marks]**

02 A student has drawn two diagrams that show the nuclei of isotopes of an element, which are shown in **Figure 1**.

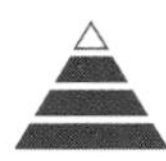

Figure 1

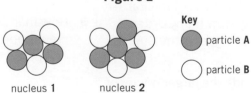

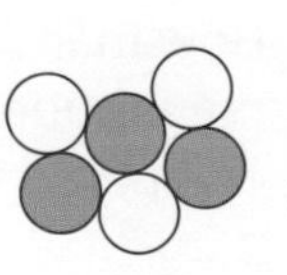

nucleus **1**

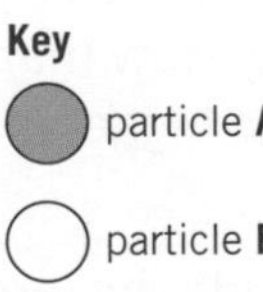

nucleus **2**

Key
particle **A**
particle **B**

02.1 Write down which particle, **A** or **B**, is a proton. Explain your answer. **[2 marks]**

02.2 Name the other type of particle. **[1 mark]**

____

Table 2

Element	Atomic number
Lithium	3
Beryllium	4
Carbon	6
Nitrogen	7

02.3 Use **Table 2** to identify the element the student drew. Explain your answer. **[2 marks]**

____

____

____

02.4 Give the complete chemical symbol of both isotopes. **[2 marks]**

____

____

02.5 Compare the charge on nucleus **1** with the charge on nucleus **2**. Justify your answer. **[2 marks]**

____

____

03 Atoms are very small.

03.1 Choose an approximate radius of an atom from the numbers in the box.

10^{15} m	10^{10} m	10^{1} m	10^{-10} m	10^{-15} m

[1 mark]

> **Exam Tip**
>
> Only select **one** answer – picking two will cause you to lose marks.

03.2 Atoms have electrons that are arranged in different energy levels. Define an 'energy level'. **[1 mark]**

03.3 **Figure 2** shows the first three energy levels in a hydrogen atom.

Normally the electron in hydrogen will be in the lowest energy level, which is level 1.

Compare what happens to the electron when an atom absorbs electromagnetic radiation with what happens when it emits electromagnetic radiation.

Use the diagram to provide examples.

[4 marks]

Figure 2

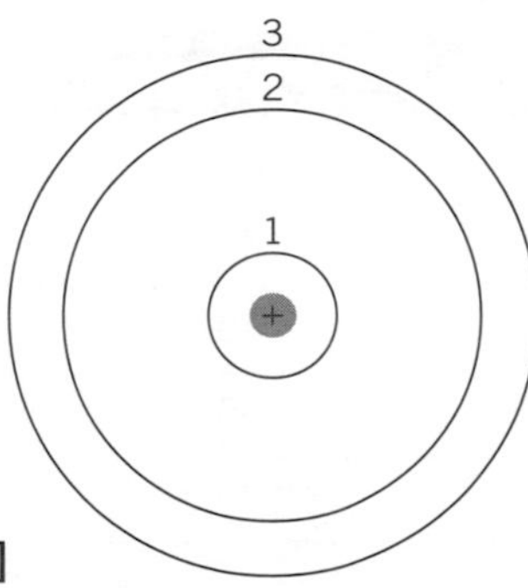

> **Exam Tip**
>
> Clearly label and lines your draw on the diagram to ensure the examiner can give you the marks for your answer.

04 A student is explaining what the symbols on the Periodic Table mean. She uses the example of the element nitrogen shown in **Figure 3**.

Figure 3

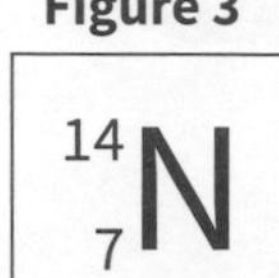

04.1 Complete the student's sentences.

You can work out the number of protons in an atom of this element by…

You can work out the number of neutrons in an atom of this element by…

You can work out the number of electrons in an atom of this element by… **[3 marks]**

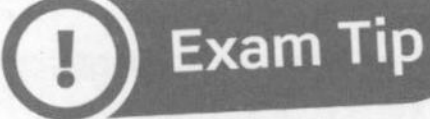

Exam Tip

Remember the mass number is the larger number and the atomic number is the smaller number.

04.2 Another student says 'The number of neutrons is always equal to the number of protons.' Is this student correct? Explain your answer. **[2 marks]**

Exam Tip

You *must* use data to back up your answer.

05 In the development of the model of the atom there have been many discoveries and different models have been developed.

- The electron was discovered in 1899 by J. J. Thompson.
- The nuclear model of the atom was proposed by Ernest Rutherford in 1911.
- The proton was named in 1920 by Rutherford.
- The neutron was discovered in 1932 by James Chadwick.

Describe how it was possible for Rutherford to develop a nuclear model before the particles that make up the nucleus had been discovered or identified. **[6 marks]**

06 A student investigates how high a ball bounces. They use a metre rule and a ball, as shown in **Figure 4**.

They drop the ball from different heights.

They measure the height of the first bounce.

Their results are shown in **Table 3**.

Figure 4

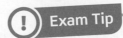

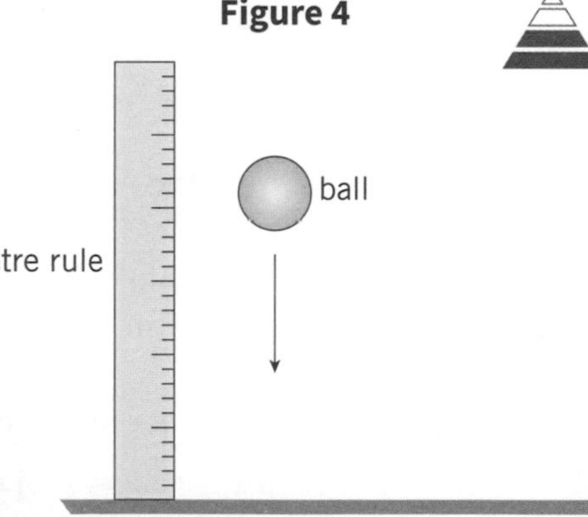

Table 3

Height of drop in cm	Height of bounce in cm
20	16
40	30
60	49
80	55
100	70

Exam Tip

This is a brilliant question to test your practical skills with. You may not be familiar with this practical, but these are exactly the sort of questions you'll get in the real exam!

06.1 Explain in terms of energy why the ball does not bounce back to the height from which it was dropped. **[2 marks]**

06.2 Plot a graph of the results. **[5 marks]**

06.3 Identify the anomalous result. **[1 mark]**

06.4 Suggest whether the bounce height is proportional to the drop height. Explain your answer. **[2 marks]**

07 **Figure 5** shows the experimental equipment that scientists working with Rutherford used to develop the nuclear model of the atom.

Figure 5

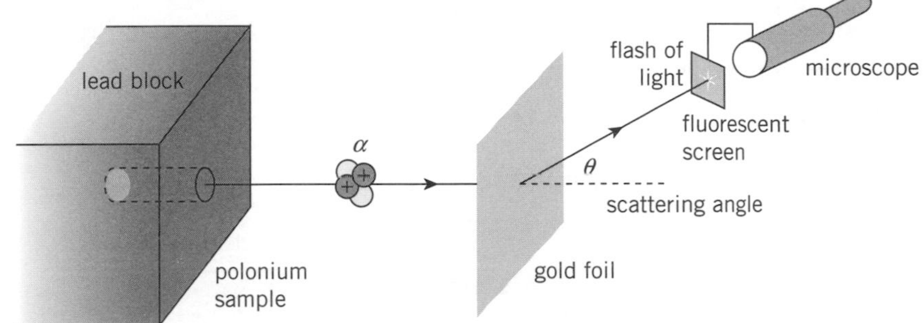

Exam Tip

To go through the foil the alpha particles must have been scattered less than 90 degrees.

Table 4 shows the number of alpha particles deflected through different angles.

Table 4

Angle of deflection	Experimental count
150	33
135	43
120	52
105	70
75	211
60	477
45	1435
30	7800
15	132 000
	Total = 142 141

07.1 Use the data in **Table 4** to show that the percentage of alpha particles that were scattered back from the foil was approximately 0.14%. **[2 marks]**

07.2 Write down the percentage of alpha particles that went through the foil. **[1 marks]**

07.3 Suggest how the data in **Table 4** may have led Rutherford to propose the nuclear model. **[2 marks]**

07.4 The alpha particle is a helium nucleus.

Write down the charge on an alpha particle. **[1 mark]**

Exam Tip

A helium nuclei has a mass of 4 and an atomic number of 2.

07.5 Suggest whether the data about the angle of deflection support the idea that the charge on the nucleus is positive or negative.

Justify your answer. **[4 marks]**

08 **Figure 6** shows transitions of electrons between different energy levels in an atom.

Figure 6

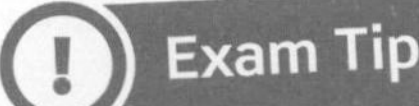

Make sure you pick arrows that are all going in the same direction – for example, either all pointing out, or all pointing in!

08.1 Write down the letters that show the transition of electrons when electromagnetic radiation is absorbed. **[2 marks]**

08.2 Write down the letters that show the transition of electrons when electromagnetic radiation is emitted. **[2 marks]**

08.3 Write down the letters that show ionisation of the atom. **[2 marks]**

08.4 Write down the charge on the atom when the atom is ionised. **[1 mark]**

08.5 The nucleus in **Figure 6** is not drawn to scale.

A student measures the diameter of the atom in a textbook and finds that it is 2 cm.

Estimate the diameter of the dot that would represent the nucleus if it was drawn to scale.

Explain your calculation.

Suggest a reason why your answer may not be feasible. **[4 marks]**

There is going to be a range of answers for this, so make sure you explain what you do fully.

09 **Figure 7** shows the plum pudding model of an atom.

Figure 7

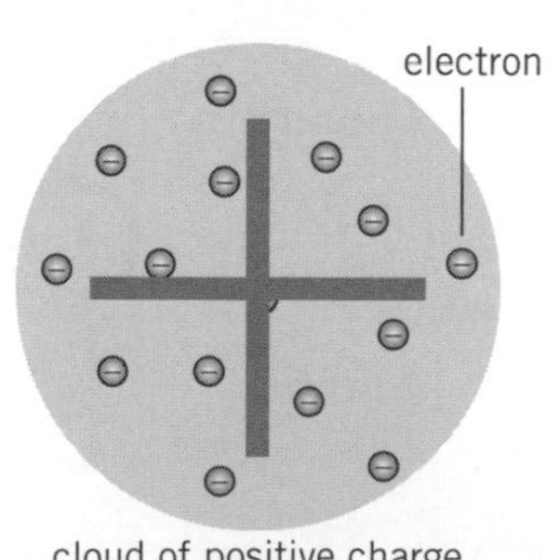

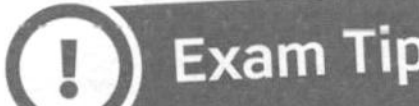

To help you answer the next question try labelling any parts you can.

09.1 Use the diagram in **Figure 8** to describe the plum pudding model. **[3 marks]**

09.2 Describe the model of the atom that the plum pudding model replaced. **[1 mark]**

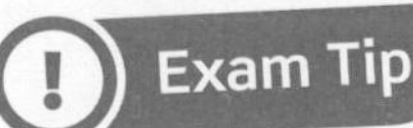

This is a great question – all the information you need to answer it is in the figure.

09.3 The plum pudding model could not explain some of the results of the alpha particle scattering experiment. Describe one of the results that the plum pudding model could **not** explain. **[2 marks]**

10 A student looks at a packet of fuse wires. They notice that the wires have different thicknesses.

10.1 Describe the action of a fuse in a circuit where an appliance is connected to the mains. **[2 marks]**

Think about the symbol for a fuse and the need for a complete circuit.

10.2 Suggest why the wires for different currents are different thicknesses. **[3 marks]**

10.3 **Table 5** shows some data relating to a particular fuse wire.

Table 5

Parameter	Fuse data
length	0.5 cm
cross-sectional area	1×10^{-6} m^2
density	7000 kg/m^3
specific heat capacity	230 J/kg °C
melting point of fuse metal	687 °C
specific latent heat	300 000 J/kg

Write down the equation that links density, mass, and volume. **[1 mark]**

10.4 Calculate the mass of the fuse wire using the data in **Table 5**. **[5 marks]**

This is a wire, so its going to be cylindrical in shape

10.5 Use the *Physics Equations Sheet* to calculate the energy needed to raise the temperature of the wire to its melting point, and then melt the fuse. **[5 marks]**

10.6 Write down the equation that links energy transferred, power, and time. **[1 mark]**

10.7 Write down the equation that links power, current, and resistance. **[1 mark]**

10.8 The resistance of the fuse wire is 1.8 Ω and the fuse melts in about 0.5 s.

Calculate the current in the wire when it melts. **[3 marks]**

11 A student finds a circuit component.

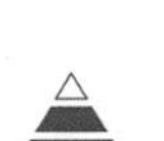

They want to make measurements to find out the relationship between the current and the potential difference for the component.

11.1 Draw a circuit diagram of the equipment that she needs to make these measurements.

Label all the components in the circuit.

Use the symbol **X** for the unknown component. **[3 marks]**

> **Exam Tip**
>
> Do not get the unknown component confused with the symbol for a bulb!

11.2 Design an experiment to find out whether or not the component is ohmic. **[4 marks]**

> **Exam Tip**
>
> Designing experiments is becoming more and more common in exams. Think about practicals you have done in class and base your method on one of those.

11.3 The student plots the graph shown in **Figure 8**.

Explain why you cannot use the gradient of the line to decide whether the component is ohmic. **[1 mark]**

Figure 8

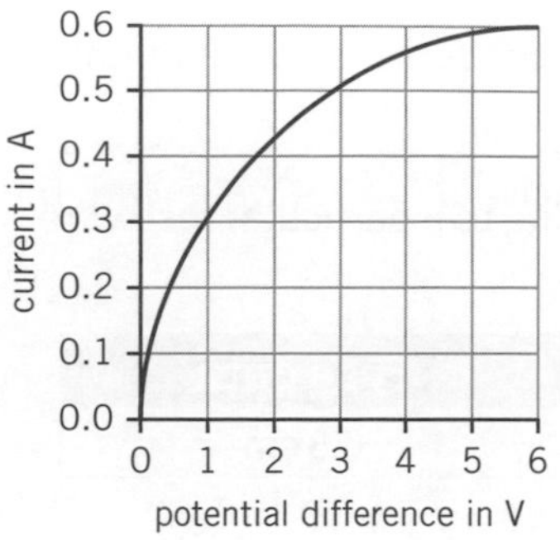

current in A (0.0–0.6) against potential difference in V (0–6)

11.4 The student reverses the battery and repeats the experiment.

Complete the graph in **Figure 8** to show what happens to the potential difference when the current is reversed. **[3 marks]**

11.5 Write down the equation that links potential difference, current, and resistance. **[1 mark]**

11.6 The resistance of the component at 2 V is 5 Ω.

Use the graph in **Figure 8** to calculate the resistance of the component at 6 V. **[2 marks]**

> **Exam Tip**
>
> Draw construction lines on the graph to show your working.

11.7 Determine whether the component is ohmic or non-ohmic.

Explain your answer. **[2 marks]**

12 A student puts a beaker of hot water on the desk. She puts a beaker of cold water next to it, as shown in **Figure 9**. The volume of water in each beaker is the same.

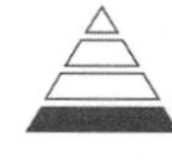

Figure 9

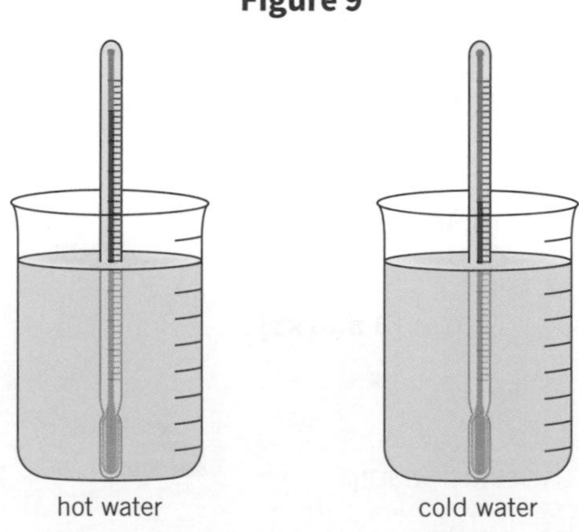

> **Exam Tip**
>
> Start this question by crossing out the answers you're not going to use, but only in pencil so you can change your mind later.

12.1 Complete the sentences below using phrases from the box.

You may need to use some phrases more than once.

greater than	the same as	lesser than

[3 marks]

The average speed of the molecules in *hot* water is _______________ the average speed of molecules in *cold* water.

The total kinetic energy of the molecules in *hot* water is ________________ the total kinetic energy of molecules in *cold* water.

The total potential energy of the molecules in *hot* water is ________________ the total potential energy of molecules in *cold* water.

12.2 Select the correct phrase to complete this sentence.

The internal energy of the molecules in hot water is **greater than / the same as / lesser than** the internal energy of molecules in cold water. **[1 mark]**

12.3 Compare the specific heat capacity of the hot water with the specific heat capacity of the cold water. **[1 mark]**

12.4 The student pours out half of the water from the beaker of hot water.

Which of the quantities (average speed, total kinetic energy, total potential energy, internal energy, or specific heat capacity) will change?

Explain your answer. **[4 marks]**

13 The nuclei of atoms can be involved in the process of nuclear fusion or nuclear fission.

Compare nuclear fission and nuclear fusion.

You should describe the two processes in your answer. **[6 marks]**

> **Exam Tip**
>
> A table can help with planning this answer, with similarities and differences on the top, and nuclear fission and fusion along the side.

14 The discovery of the electron led to the development of the plum pudding model of the atom by J.J. Thompson.

14.1 Describe the plum pudding model of the atom. **[2 marks]**

14.2 Explain why the discovery of the electron was important in the development of the model of the atom. **[2 marks]**

14.3 The plum pudding model was replaced by the nuclear model due to evidence from the alpha particle scattering experiment.

Suggest what a scientist would have expected to see in the alpha particle experiment if the plum pudding model was correct. **[1 mark]**

> **Exam Tip**
>
> Practice at predicting results based on a model is great preparation for the exam!

14.4 Use your answer to **14.3** to explain why the alpha particle experiment resulted in a change to the model of the atom. **[2 marks]**

14.5 The results of the experiment were published in 1909, but the nuclear model was not proposed until 1911.

Suggest **one** reason why the model was proposed after the results were published. **[1 mark]**

Knowledge

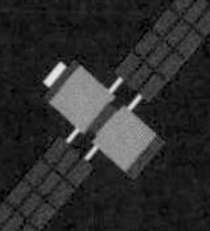
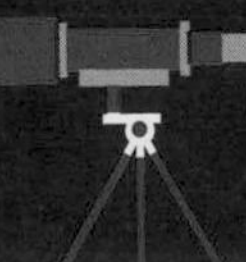

P8 Radiation

Radioactive decay

Radioactive decay is when nuclear radiation is emitted by unstable atomic nuclei so that they become more stable. It is a *random* process. This radiation can knock electrons out of atoms in a process called **ionisation**.

Type of radiation	Change in the nucleus	Ionising power	Range in air	Stopped by	Decay equation
α alpha particle (two protons and two neutrons)	nucleus loses two protons and two neutrons	highest ionising power	travels a few centimetres in air	stopped by a sheet of paper	${}^{A}_{Z}X \rightarrow {}^{(A-4)}_{(Z-2)}Y + {}^{4}_{2}\alpha$
β beta particle (fast-moving electron)	a neutron changes into a proton and an electron	high ionising power	travels ≈ 1 m in air	stopped by a few millimeters of aluminium	${}^{A}_{Z}X \rightarrow {}^{A}_{(Z+1)}Y + {}^{0}_{-1}\beta$
γ gamma radiation (short-wavelength, high-frequency EM radiation)	some energy is transferred away from the nucleus	low ionising power	virtually unlimited range in air	stopped by several centimetres of thick lead or metres of concrete	${}^{A}_{Z}X \rightarrow {}^{A}_{Z}X + {}^{0}_{0}\gamma$

Activity and count rate

The **activity** of a radioactive source is the rate of decay of an unstable nucleus, measured in becquerel (Bq).

1 Bq = 1 decay per second

Detectors (e.g., **Geiger-Muller tubes**) record a **count rate** (number of decays detected per second).

$$\text{count rate after } n \text{ half-lives} = \frac{\text{initial count rate}}{2^n}$$

Half-life

The **half-life** of a radioactive source is the time

- for half the number of unstable nuclei in a sample to decay
- for the count rate or activity of a source to halve.

The half-life of a source can be found from a graph of its count rate or activity against time.

To find the reduction in activity after a given number of half-lives:

1 calculate the activity after each half-life

2 subtract the final activity from the original activity.

Net decline can be given as a ratio:

$$\text{net decline} = \frac{\text{reduction in activity}}{\text{original activity}}$$

Irradiation versus contamination

irradiation	when an object is exposed to nuclear radiation	cause harm through ionisation	prevented by shielding, removing, or moving away from the source of radiation
contamination	when atoms of a radioactive material are on or in an object		object remains exposed to radiation as long as it is contaminated contamination can be very difficult to remove

Protection against irradiation and contamination

You can protect against irradiation and contamination by:

- maintaining a distance from the radiation source
- limiting time near the source
- shielding from the radiation.

Studies on the effects of radiation should be published, shared with other scientists, and checked by **peer review** as they are important for human health.

Ionising radiation

Living cells can be damaged or killed by ionising radiation.

The risk depends on the half-life of the source and the type of radiation.

Alpha radiation is very dangerous inside the body because it affects all the surrounding tissue. Outside the body it only affects the skin and eyes because it cannot penetrate further.

Beta and gamma radiation are dangerous outside and inside the body because they can penetrate into tissues.

Radiation dose

Radiation dose, measured in sievert (Sv), measures the health risk of exposure to radiation. It depends on the type and amount of radiation.

Background radiation

Background radiation is radiation that is around us all the time. It comes from:

- natural sources like rocks and cosmic rays
- nuclear weapons and nuclear accidents.

Background radiation is always present but the levels are higher in some locations and in some jobs.

Nuclear radiation in medicine

Exploration of internal organs

Gamma-emitting **tracers** are injected or swallowed by a patient. Gamma cameras can then create an image showing where the tracer has gone.

The half-life of the tracer must be short enough so that most of the nuclei will decay shortly after the image is taken to limit the patient's radiation dose (normally about six hours).

Control or destruction of unwanted tissue

1. Narrow beams of gamma radiation can be focused on tumour cells to destroy them. Gamma is used because it can penetrate tumours from outside the body.
2. Beta- or gamma-emitting implants can be surgically placed inside (or next to) tumours. Their half-lives must be long enough to be effective, but short enough that it does not continue to irradiate the patient after treatment.

Nuclear fission

Nuclear fission is when a large unstable nucleus absorbs an extra neutron and splits into two smaller nuclei of roughly equal size.

During nuclear fission:

- gamma radiation is emitted and energy is released
- two or three neutrons are emitted that can go on to cause a **chain reaction**.

The chain reaction in a power station reactor is controlled by absorbing neutrons.

Nuclear explosions are uncontrolled chain reactions.

On rare occasions an unstable nucleus splits apart without absorbing a neutron. This is called **spontaneous fission**.

Nuclear fusion

Nuclear fusion is when two light nuclei join to make a heavier one.

Some of the mass is converted to energy and transferred as radiation.

Key terms Make sure you can write a definition for these key terms.

alpha activity background radiation beta chain reaction contamination count rate fission fusion gamma Geiger-Muller tube half-life ionisation irradiation net decline peer review radiation dose radioactive decay spontaneous fission tracer

Retrieval

Learn the answers to the questions below then cover the answers column with a piece of paper and write as many as you can. Check and repeat.

	P8 questions		Answers
1	What are the three types of nuclear radiation?	Put paper here	alpha, beta, and gamma
2	What is gamma γ radiation?		electromagnetic radiation from the nucleus
3	Which type of nuclear radiation is the most ionising?		alpha
4	What is the range in air of alpha, beta, and gamma radiation?	Put paper here	a few cm, 1 m, and unlimited, respectively
5	Which materials can stop alpha, beta, and gamma radiation?		sheet of paper, thin aluminium sheet, and thick lead/concrete, respectively
6	Which type of nuclear radiation does not cause a change in the structure of the nucleus when it is emitted?	Put paper here	gamma
7	What are the equation symbols for alpha and beta particles?		${}^{4}_{2}\alpha$ and ${}^{0}_{-1}\beta$
8	What is radioactive activity?	Put paper here	the rate at which a source of unstable nuclei decays
9	What unit is used to measure the activity of a radioactive source?		becquerel (Bq)
10	What is 'count-rate'?	Put paper here	number of decays recorded each second (by a detector, e.g., Geiger-Muller tube)
11	What is meant by the half-life of a radioactive source?		time taken for half the unstable nuclei to decay or the time taken for the count rate to halve
12	What is irradiation?	Put paper here	exposing an object to nuclear radiation
13	What is radioactive contamination?		unwanted presence of substances containing radioactive atoms on or in other materials
14	Where does background radiation come from?	Put paper here	rocks, cosmic rays, fallout from nuclear weapons testing, nuclear accidents
15	Why are gamma-emitting sources used for medical tracers and imaging?		gamma rays pass through the body without causing much damage to cells
16	What is nuclear fusion?	Put paper here	when two light nuclei join to make a heavier one
17	What is nuclear fission?		the splitting of a large and unstable nucleus into two smaller nuclei
18	How does nuclear fission occur?		an unstable nucleus absorbs a neutron, it splits into two smaller nuclei, and emits two or three neutrons plus gamma rays

Now go back and use the questions below to check your knowledge from previous chapters.

Previous questions | Answers

	Previous questions	Answers
1	Describe the plum pudding model of the atom.	sphere of positive charge with negative electrons embedded in it
2	Give two reasons why the temperature of a gas increases if it is compressed quickly.	the force applied to compress the gas results in work being done to the gas, and the energy gained by the gas is not transferred quickly enough to the surroundings
3	What is the radius of an atom?	around 1×10^{-10} m
4	Why does a graph showing the change in temperature as a substance cools have a flat section when the substance is changing state?	the energy transferred during a change in state causes a change in the internal energy of the substance
5	What two things does energy transfer to an appliance depend on?	power of appliance, time it is switched on for
6	What are step-up transformers used for in the National Grid?	increase the p.d. from the power station to the transmission cables
7	What are the main advantages of using solar energy?	can be used in remote places, no polluting gases, no waste products, very low running cost

Put paper here

Put paper here

Put paper here

Maths Skills

Practise your maths skills using the worked example and practice questions below.

Ratios, fractions, percentages	Worked example	Practice
A **ratio** is a way of comparing the size of two quantities. For example, a ratio of 2:4 of radioactive atoms to non radioactive atoms in a sample means for every 2 radioactive atoms, there are 4 non radioactive atoms. A ratio can be simplified by dividing both numbers by their highest common factor. A **fraction** can be a way of expressing part of a whole number, or a way of writing one number divided by another in an equation. To find fractions from a ratio, each number in the ratio can be a numerator, and the denominator is the sum of both numbers. For example, if the ratio of apples to oranges is 2:3, the fraction of apples is $= \frac{2}{2+3} = \frac{2}{5}$ A **percentage** is a number expressed as a fraction of 100. For example, 45% $= \frac{45}{100}$ To find one number as a percentage of another divide the first number by the second and multiply by 100.	A sample has a ratio of 8:20 radioactive atoms to non radioactive atoms. Simplify this ratio, and find the fraction that are radioactive atoms. **Answer:** Find the greatest common factor for 8 and 12 = 4. Divide both sides of the ratio by 4 = 2:5. Fraction of radioactive atoms: $= \frac{2}{2+5} = \frac{2}{7}$ The resistance of a thermistor changes from 250 Ω to 175 ohms when it is heated. Calculate the percentage change in its resistance. **Answer:** Calculate the change in resistance: 250 – 175 = 75 Divide the change by the original value of the resistance: $\frac{75}{250} = 0.3 \times 100 = 30\,\%$	1 A sample has 40 radioactive atoms for every 120 non radioactive atoms. Write this as a ratio in its simplest form. 2 In the above example, what fraction of atoms are not radioactive? 3 In the above example, what percentage of atoms are radioactive?

Practice

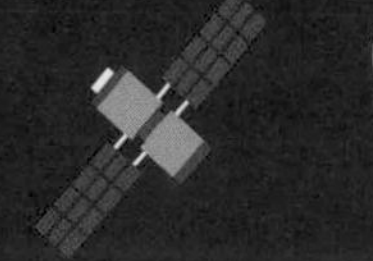
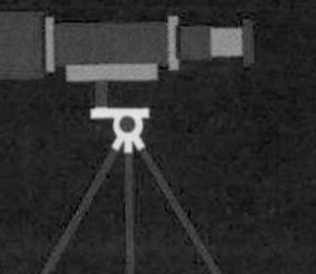

Exam-style questions

01 One of the uses of radioactive materials is in smoke detectors.

The isotope inside the smoke detector produces alpha radiation.

When there is smoke inside the detector, the smoke stops a current flowing in a circuit. This causes an alarm to go off.

01.1 Explain why the source needs to produce alpha radiation, and not beta or gamma radiation. **[2 marks]**

____

____

____

01.2 The decay of the isotope used in the smoke detector can be shown by the equation:

$$^{241}_{95}\text{Am} \rightarrow {}^{4}_{2}\alpha + {}^{\square}_{\square}\text{Np}$$

The equation shows how an americium nucleus decays into a neptunium nucleus.

Calculate the atomic number and atomic mass of the neptunium.

Show your working. **[4 marks]**

> **! Exam Tip**
>
> The maths here may look hard but its not. Take it one line at a time.

Atomic number = ______________

Atomic mass = ______________

01.3 Explain why americium decays into a different element. **[2 marks]**

____

____

____

> **! Exam Tip**
>
> Think about what is the one thing that all atoms and ions of an element have in common. What makes one element different from another?

02 One atom in 10^{10} atoms of carbon is an atom of a carbon-14 isotope.

Figure 1 shows the count rate of carbon-14 against time.

Figure 1

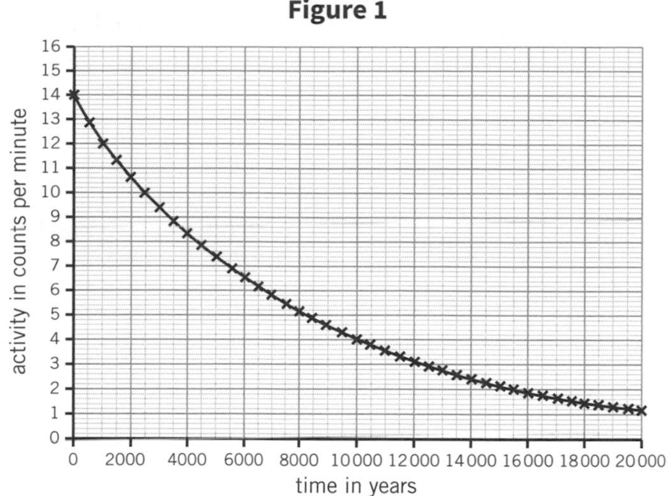

02.1 Deduce the half-life of carbon-14. **[1 mark]**

____

02.2 An archaeologist finds a fragment of a wooden spear. The spear contains carbon.

The fragment has an activity of 5 counts per minute.

Use **Figure 1** to deduce the age of the spear. **[1 mark]**

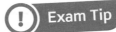

______________ years

> **Exam Tip**
>
> You can simply read the answer off **Figure 1** for both **02.1** and **02.2**.

02.3 The last ice age ended around 11 000 years ago.

Is the spear old enough to have been used during the last ice age?
Give a reason for your answer. **[2 marks]**

____

____

03 A teacher demonstrates how to measure the activity of a radioactive material.

03.1 Which **two** statements about activity are correct?
Choose **two** answers. **[2 marks]**

The activity of a sample is the number of particles it emits.

The activity of a sample is measured in becquerels (Bq).

The activity of a sample is the amount of radiation it emits.

The activity of a sample is the number of decays recorded per second.

> **Exam Tip**
>
> As you read through the options:
> - cross out any you know are wrong
> - put a question mark next to any your unsure about
> - put a tick next to the ones you are confident are correct.
>
> That will only leave you a few to pick from.

03.2 Suggest a detector that the teacher could use to measure the activity. **[1 mark]**

03.3 The radiation emitted by a nucleus can

- be a particle or an electromagnetic wave
- be charged or have no charge.

Complete **Table 1** by ticking the correct boxes in the second and third columns. **[2 marks]**

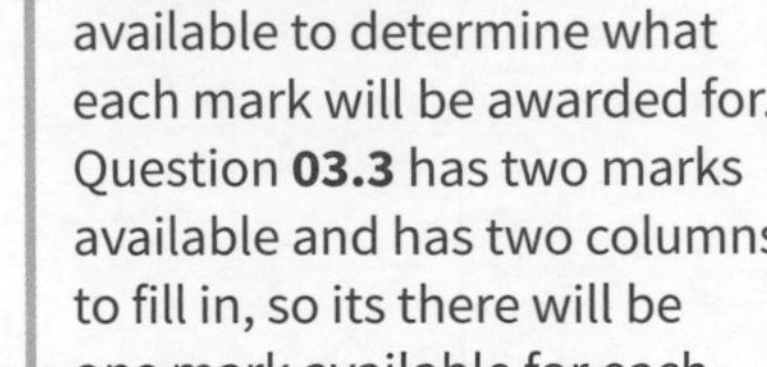

You can use the the marks available to determine what each mark will be awarded for. Question **03.3** has two marks available and has two columns to fill in, so its there will be one mark available for each column not for each tick.

Table 1

Type of radiation	Is a particle	Has no charge
alpha		
beta		
gamma		
neutron		

03.4 A teacher uses dice to demonstrate what happens when radioactive material decays. Suggest **one** reason why throwing dice is a good model for radioactive decay. Explain your answer. **[2 marks]**

04 Different types of radiation can travel different distances through the air.

04.1 Complete **Table 2** with the words alpha, beta, and gamma. **[2 marks]**

Table 2

Type	Range in air
	>3 m
	1 m
	<10 cm

04.2 A student says:

'*There is a link between the ionising power of radiation and how far they go in air. If they do not go as far, that means that they are not as ionising.*'

Do you agree with the student? Explain your answer. **[3 marks]**

04.3 A teacher has a source that emits all three types of radiation. The activity of the source is 35 Bq. The teacher puts a sheet of aluminium between the source and the detector. The activity recorded is lower. Explain why the activity is lower, but radiation can still be detected. **[2 marks]**

05 Caesium-137, $^{137}_{55}\text{Cs}$, has a half-life 30 years.

05.1 Determine the mass of caesium-137 that remains in a 24 g sample after 90 years. **[3 marks]**

05.2 Complete the equation for the decay of caesium-137 when it emits a beta particle. **[2 marks]**

$$^{137}_{55}\text{Cs} \rightarrow {}^{137}_{\square}\text{Ba} + \square$$

05.3 Another isotope is caesium-134. Caesium-134 was emitted during an explosion at the Chernobyl nuclear reactor in Ukraine. Following the explosion, caesium-134 isotopes were found in fields in Wales. Sheep were farmed in these fields. Explain why the presence of radioactive material on the grass produces a hazard for sheep. **[2 marks]**

06 A teacher is modelling a chain reaction. She uses long matches in a tray of sand. She arranges the matches so that when she lights the first match the other matches are ignited.

06.1 Describe how this demonstration is a model for a fission chain reaction. **[3 marks]**

06.2 Suggest **two** strengths and **two** limitations of the model. **[4 marks]**

Exam Tip

Clearly indicate which points you give are the strengths and which are the limitations.

07 **Table 3** shows the average percentage of background radiation that a person gets from different sources of background radiation.

Table 3

Source	Percentage
radon gas	48.0
rocks and soil	15.0
inside our bodies	13.0
medical X-rays and γ-rays	12.0
cosmic rays	10.0
nuclear fallout	0.8
air travel	0.4
nuclear waste	0.2
industry	0.2
luminous watches	0.1

07.1 Suggest what 'inside our bodies' might mean in terms of a source of radiation. **[1 mark]**

07.2 Describe where cosmic rays come from. **[1 mark]**

07.3 Radiation dose is measured in millisieverts (mSv). If you live in the UK your average annual dose of radiation is 2.7 mSv. Give **one** reason why someone might have an average annual dose that was significantly higher than 2.7 mSv. **[1 mark]**

07.4 Calculate the dose in mSv that the average person will get from radon gas. **[2 marks]**

07.5 Suggest the type of graph or chart a student could plot to show the data in **Table 3**. Give reasons for your answer. **[2 marks]**

08 When a nucleus decays the mass of the nucleus might, or might not, change.

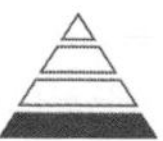

08.1 Write down a type of radiation that does **not** change the mass of the nucleus when it is emitted. **[1 mark]**

08.2 An equation that shows the decay of bismouth-214 is

$$^{214}_{83}\text{Bi} \rightarrow ^{214}_{84}\text{Po} + \textbf{X}$$

Name particle **X**. Give a description of this particle. **[2 marks]**

08.3 Explain why the mass of the nucleus does not change. **[2 marks]**

08.4 Describe how the decay equation shows what happens to the charge on the nucleus when particle **X** is emitted. Explain your answer. **[2 marks]**

09 Technetium isotopes are used for medical imaging. Doctors inject a patient with a very small amount of the technetium isotope, which is taken up by an organ of the body. The doctor looks at the emitted radiation on a gamma camera. **Table 4** shows three isotopes of technetium, the radiation that they emit, and their half-lives.

Table 4

Name	Type of emitter	Half life
technetium-95	gamma	61 days
technetium-99	beta	2.1×10^5 years
technetium-99m	gamma	6 hours

Suggest which isotope should be used as a tracer. Justify your answer. Suggest the consequences of using one of the other isotopes listed in **Table 4**. **[6 marks]**

Exam Tip

You need to consider a few things for this question.

- Which one will be safest for the patient?
- Which one will last long enough for the doctors to be able to record the results?
- Which one will not cause long term damage?

10 A student collects data on the activity of a sample of radioactive material. The data are show in **Table 5** and **Figure 2**.

Table 5

Time in s	Count rate in Bq
0	42.6
20	35.2
40	24.9
60	20.9
80	17.7
100	15.2
120	10.7
140	8.7
160	4.5
180	6.1
200	4.3
220	3.7

Figure 2

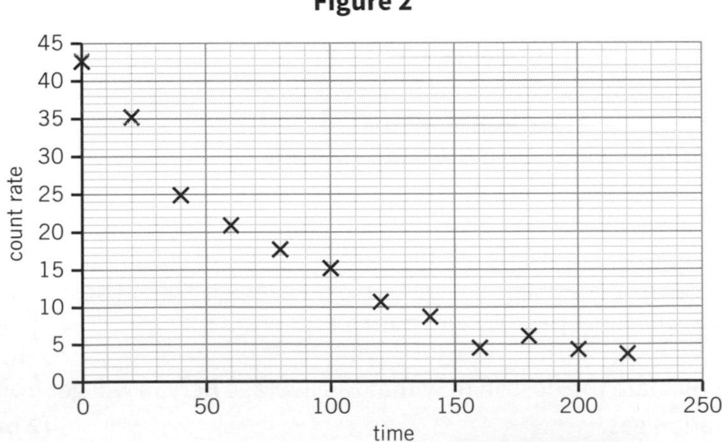

10.1 Draw a line of best fit on **Figure 2**. **[1 mark]**

10.2 Use **Figure 2** to find the half-life of the isotope. **[3 marks]**

10.3 A student says that the data is wrong because there are some points that are not on the line of best fit. Do you agree? Justify your answer. **[2 marks]**

Exam Tip

A lines of best fit is *not* a dot-to-dot. You do not have to include every point in your line. It should show the trend of the data. Use a ruler to see if all the points will fit on a straight line. If they don't, draw a curved line.

10.4 Identify **one** thing that the student has done incorrectly when producing **Figure 2**. Assume that all the data points are correct. **[1 mark]**

Exam Tip

Look at the data carefully and compare it to the points plotted on the graph.

11 Strawberry fruits have microorganisms on them that cause the strawberries to decay. If the strawberries are irradiated, the radiation kills the microorganisms. Some people do not like to eat irradiated strawberries because they think that they themselves will become contaminated.

11.1 Describe the difference between contamination and irradiation in this context. **[2 marks]**

Exam Tip

The key part of this question is 'in this context'. You must refer to the context given, otherwise you won't gain full marks even if correctly describe the difference between contamination and irradiation.

11.2 Explain why the hazard due to radiation is low when eating irradiated strawberries. **[1 mark]**

11.3 There are regulations that cover processes involving radioactive materials. The data that are used to produce the regulations come from reports published in peer-reviewed journals. Describe the process of peer review. Explain why it is very important for regulations covering these processes to be based on peer-reviewed articles. **[3 marks]**

12 A student uses a heater to find the specific heat capacity of a liquid. She connects a heater to an energy meter connected to a datalogger. She puts a temperature probe into a beaker of liquid and uses the data logger to record the temperature of the liquid over time. **Table 6** shows the student's data.

Table 6

Time in s	Temperature in °C
0	25.0
16	32.5
32	39.0
48	45.5
64	54.5
80	61.0
96	69.5
112	75.0
128	84.0
144	90.5
160	97.0
176	98.5
192	98.5
208	98.5

12.1 Explain the trend shown by **Table 6** in terms of energy. **[4 marks]**

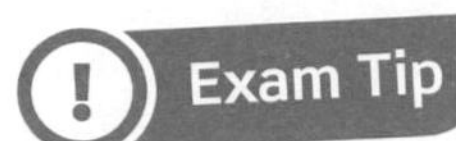

Exam Tip

This is an explain question, not a describe question. Therefore you have to identify the trend shown, then explain why that trend happens. Make sure you reference energy, as stated in the question.

12.2 The student looks up a graph of energy transferred to the liquid against temperature of the liquid in a textbook (**Figure 3**).

Figure 3

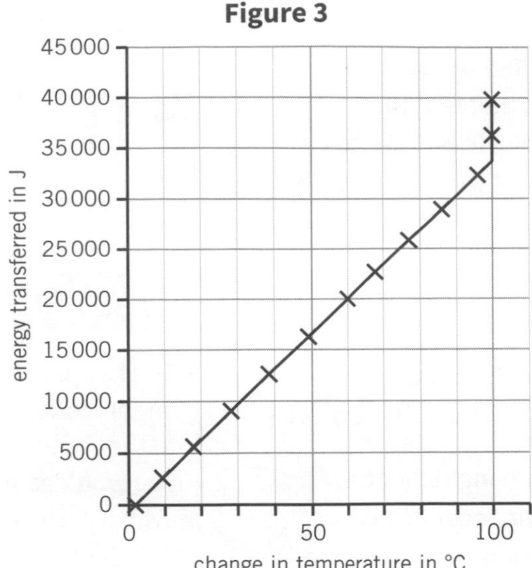

Describe the trend shown by **Figure 3**. **[2 marks]**

12.3 Use **Figure 3** to calculate the specific heat capacity of the liquid. Give your answer to two significant figures. Assume the mass of the liquid is 100 g. Use the correct equation from the *Physics Equations Sheet*. **[4 marks]**

12.4 Suggest why it is better to calculate the specific heat capacity at a change in temperature of 80 °C instead of at 50 °C. **[1 mark]**

> **Exam Tip**
>
> If you are unsure what equation to use, find the equation that includes what you've been asked to calculate. If you are still unsure, read through the whole question and write down all the variables you know. Remember to include any you may have calculated in an earlier question. Then look for the equation that includes these.

> **Exam Tip**
>
> Watch out for units in **12.3**.

13 A student sets up an experiment to compare the potential difference across components in series and parallel circuits. The circuit diagrams are shown in **Figure 4**.

Figure 4

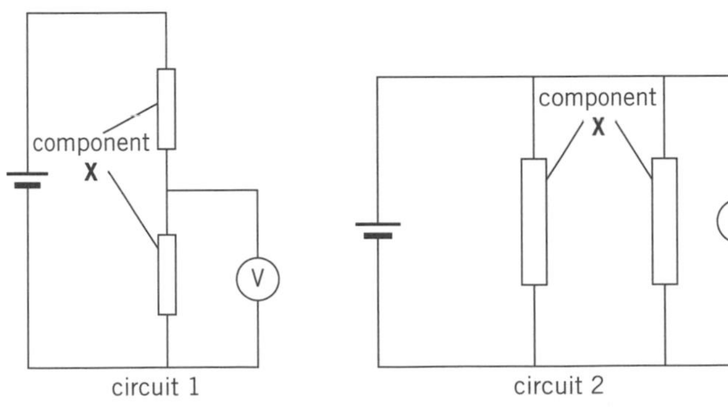

13.1 Identify component **X**. **[1 mark]**

13.2 The cell in each circuit has a potential difference of 6 V. Compare the readings on the voltmeters in the two circuits.

Explain your answer. **[4 marks]**

> **Exam Tip**
>
> You need to identify the circuit symbol.

13.3 The student replaces all of the components labelled **X** with bulbs. The bulbs are identical. Suggest what happens to the readings on the voltmeters. Explain your answer. **[2 marks]**

14 In some parts of the world there are many hours of sunshine every day. People use photovoltaic (PV) cells on their houses to generate electricity.

Figure 5 shows the PV output of 20 homes over one day and the amount of electricity used in the homes. The homes are also connected to the National Grid.

Figure 5

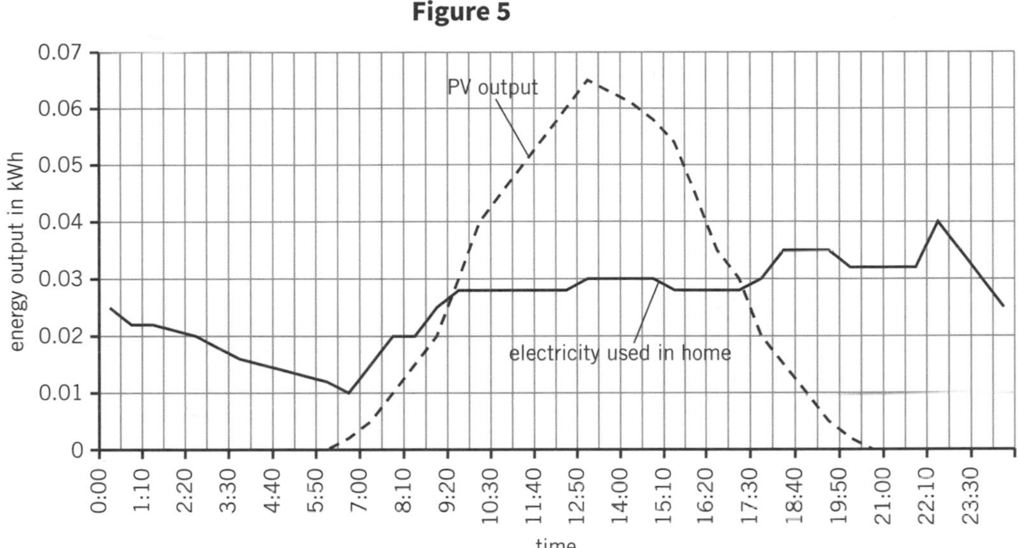

14.1 Explain why photovoltaic cells use a renewable energy resource. **[2 marks]**

14.2 Define the term National Grid. **[1 mark]**

14.3 There is a difference between the output of the PV cells and the energy that is needed in the home. If the output of the PV cells is greater than needed, the homeowner can sell the electricity back to the National Grid.

Identify the time period during the day when the output of the PV cells is greater than is needed by the houses. **[1 mark]**

14.4 Describe what happens when the output of the PV cells is less than the energy needed in the home. **[1 mark]**

14.5 If it is cloudy, the output of the PV cells drops below that needed to power a house.

Suggest **two** other reasons why a person might choose *not* to have PV cells on their house. **[2 marks]**

Knowledge

P9 Forces

Scalars and vectors

Scalar quantities only have a magnitude (e.g., distance and speed).

Vector quantities have a magnitude *and* a direction (e.g., velocity and force).

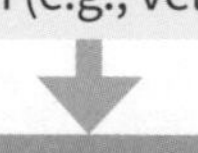

Forces

A **force** can be a push or pull on an object caused by an interaction with another object. Forces are vector quantities.

Contact forces occur when two objects are touching each other. For example, friction, air-resistance, tension, and normal contact force.

Non-contact forces act at a distance (without the two objects touching). For example, gravitational force, electrostatic force, and magnetic force.

When an object exerts a force on another object, it will experience an *equal and opposite* force.

Gravity

The force of **gravity** close to the Earth is due to the planet's **gravitational field strength**.

Weight is the force acting on an object due to gravity.

The weight of an object

- can be considered to act at the object's **centre of mass**
- can be measured using a calibrated spring-balance (newtonmeter).

weight (N) = mass (kg) × gravitational field strength (N/kg)

$$W = m\,g$$

Weight and mass are directly proportional to each other, which can be written as $W \propto m$, so as the mass of an object doubles, its weight doubles.

Resultant forces

If two or more forces act on an object along the same line, their effect is the same as if they were replaced with a single **resultant force**. The resultant force is

- the sum of the magnitudes of the forces if they act in the same direction
- the difference between the magnitudes of the forces if they act in opposite directions.

If the resultant force on an object is zero, the forces are said to be **balanced**.

If the forces do not act along the same line, the resultant of two forces can be found by making a scale drawing using a ruler and a protractor.

Drawing forces

Free body diagrams use arrows to show all of the forces acting on a single object. For example:

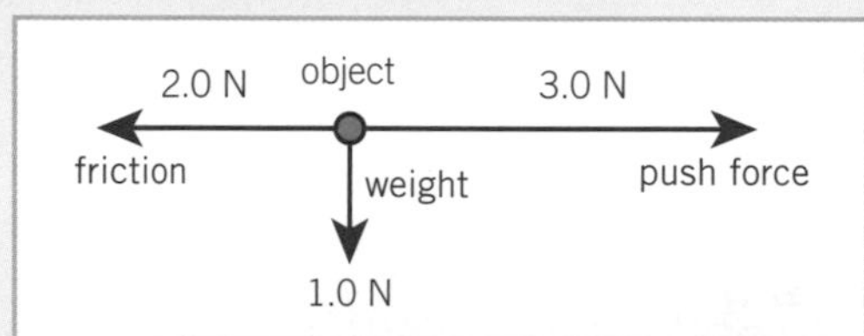

A dot or circle represents the object, with the forces drawn as arrows:

- the arrow length represents the magnitude of the force
- the arrow direction shows the direction of the force.

Scale drawings

Scale drawings can be used to find the resultant of two forces which are not acting along the same line.

The forces are drawn end to end. The resultant can then be drawn between the two ends, forming a triangle:

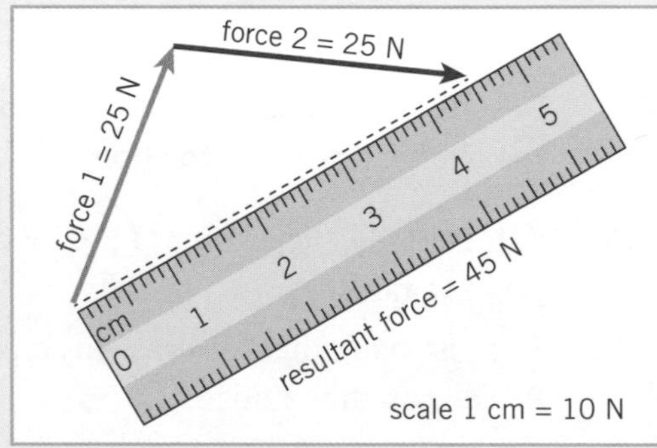

Key terms

Make sure you can write a definition for these key terms.

contact force, deformation, elastic, free body diagram, force, gear, gravitational field strength, gravity, inelastic, lever, limit of proportionality, moment, non-contact force, resultant, scalar, vector, weight

Deformation

Deformation is a change in the shape of an object caused by stretching, squashing (compressing), bending, or twisting.

More than one force has to act on a stationary object to deform it, otherwise the force would make it move.

Elastic deformation – the object can go back to its original shape and size when the forces are removed.

Inelastic deformation – the object does not go back to its original shape or size when the forces are removed.

Graphs of force against extension for elastic objects

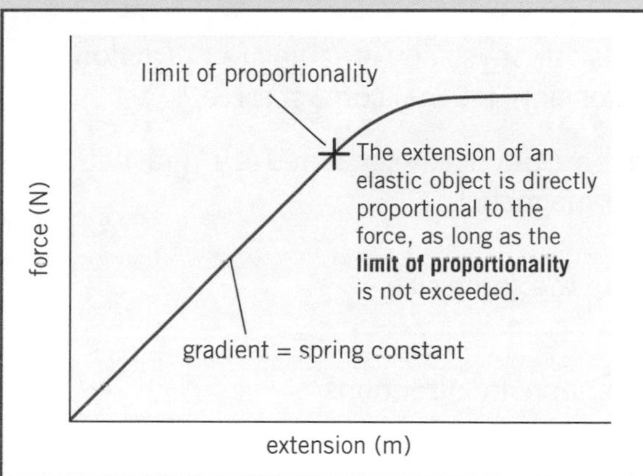

limit of proportionality
The extension of an elastic object is directly proportional to the force, as long as the **limit of proportionality** is not exceeded.
force (N)
gradient = spring constant
extension (m)

The spring constant can be calculated using the equation:

force applied (N) = spring constant (N/m) × extension (m)

$$F = k\,e$$

This relationship also applies to compressing an object, where *e* would be compression instead of extension.

Elastic potential energy

A force that stretches or compresses an object does work on it, causing energy to be transferred to the object's elastic potential store.

The elastic potential energy stored in an elastically stretched or compressed spring can be calculated using:

elastic potential energy (J) = $\frac{1}{2}$ × spring constant (N/m) × (extension)^2 (m^2)

$$E_e = \frac{1}{2}\,k\,e^2$$

Revision tip

Remember your maths skills here.

BODMAS: it is only the value for extension that is squared, not the whole answer.

Resolving forces

A single force can always be resolved (split) into two component forces at right angles to each other:

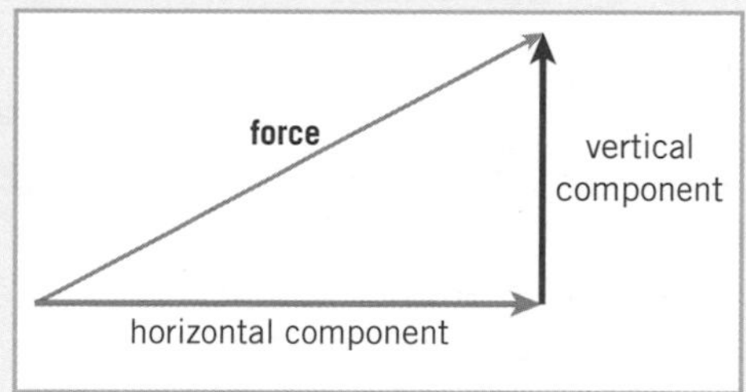

The two component forces added together give the same effect as the single force.

Moments

A force or system of forces can cause an object to rotate.

The turning effect of a force is called the **moment** of the force, and its size can be calculated using the equation:

moment of a force (Nm) = force (N) × distance (m)

$$M = F\,d$$

If an object is balanced, the sum of the clockwise moments equals the sum of the anticlockwise moments.

Levers and gears

Levers and **gears** can be used to increase the moment of a force, making it easier to lift or rotate an object.

If a small gear drives a large gear, the moment of the applied force is *increased* but the large gear moves slower (and vice versa).

A lever allows a large moment of force to be produced by allowing force to be applied further from the pivot.

Learn the answers to the questions below then cover the answers column with a piece of paper and write as many as you can. Check and repeat.

	P9 questions		Answers
1	What is a scalar quantity?	Put paper here	only has a size (magnitude)
2	What is a vector quantity?		has both a size and direction
3	What is a force?		a push or pull that acts on an object due to the interaction with another object
4	Is force a vector or scalar quantity?	Put paper here	vector
5	What is a contact force?		when objects are physically touching (e.g., friction, air-resistance, tension, normal contact force)
6	What is a non-contact force?	Put paper here	when objects are physically separated (e.g., gravitational, electrostatic, magnetic)
7	What is the same about the interaction pair of forces when two objects interact with each other?		the forces are the same size
8	What is different about the interaction pair of forces when two objects interact with each other?	Put paper here	forces are in opposite directions
9	What is the size of the resultant force on an object if the forces on it are balanced?		zero
10	What is the name for the force acting on an object due to gravity?		weight
11	What instrument can be used to measure the weight of an object?	Put paper here	calibrated spring-balance (newtonmeter)
12	What is the centre of mass?		the point through which the weight of an object can be considered to act
13	What is elastic deformation?	Put paper here	an object can go back to its original shape and size when deforming forces are removed
14	What is inelastic deformation?		an object does not go back to its original shape and size when deforming forces are removed
15	How do you find the spring constant from a force–extension graph of a spring?	Put paper here	find the gradient of the straight line section
16	What is the turning effect of a force called?		a moment
17	What can you say about clockwise and anticlockwise moments on a balanced object?	Put paper here	sum of all the clockwise moments about any point = sum of all the anticlockwise moments about that point
18	How does a lever reduce the amount of force needed to create a particular sized moment?		by increasing the distance from the pivot
19	What happens to the moment of a force when a small gear drives a large gear?		moment is increased

Now go back and use the questions below to check your knowledge from previous chapters.

Previous questions | Answers

	Previous questions	Answers
1	What are the equation symbols for alpha and beta particles?	${}^{4}_{2}\alpha$ and ${}^{0}_{-1}\beta$
2	Which particle do atoms of the same element always have the same number of?	protons
3	Explain why the pressure of a fixed mass of gas decreases if the volume is increased and kept at constant temperature.	the distance the particles travel between each impact with a wall of the container is greater, so the number of impacts per second decreases, so the total force of the impacts decreases
4	Why are gamma-emitting sources used for medical tracers and imaging?	gamma rays pass through the body without causing much damage to cells
5	What is 'count-rate'?	number of decays recorded each second (by a detector, e.g., Geiger-Muller tube)

Put paper here

Put paper here

Required Practical

Extension of a spring

In this practical you measure the extension of a spring as different forces are applied to it.

To be accurate and precise you need to:

- measure extension using a pointer directed at the same position on the spring each time
- ensure that the ruler is positioned so that it is parallel to the spring
- make measurements by looking in a direction perpendicular to the ruler
- convert mass to weight (force) if necessary.
- use a measurement of zero force = zero extension
- use the equation

$$\text{gradient} = \frac{1}{\text{spring constant}}$$

for a graph of force against extension

Worked example

A student records the following measurements for a spring.

Force in N						
Mass in kg	0	0.1	0.2	0.3	0.4	0.5
Length in cm	5.0	9.2	13.1	17.4	21	25
Extension in cm						

1 Calculate the spring extensions and forces.

weight (= force) = mass × gravitational field strength (= 10 N/kg)

extension = length – 5 (original length of spring)

Force in N	0	1	2	3	4	5
Extension in cm	0	4.2	8.1	12.4	16	20

2 Plot a graph of the results. Calculate the spring constant from your graph and give the unit.

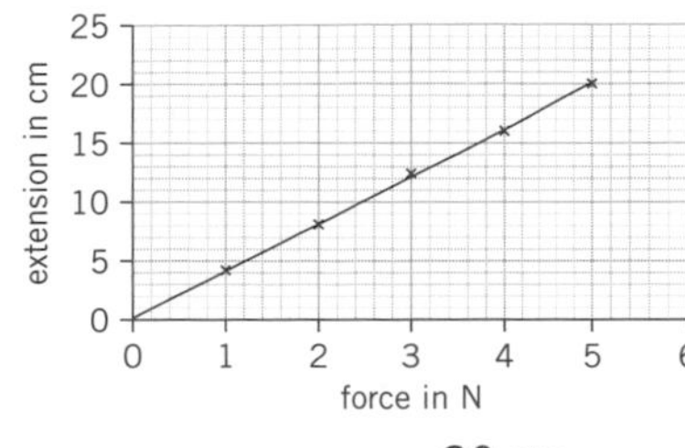

$$\text{gradient} = \frac{20\,\text{cm}}{5\,\text{N}} = 4$$

$$\text{gradient} = \frac{1}{\text{spring constant}}$$

$$\text{so spring constant} = \frac{1}{\text{gradient}} = \frac{1}{4} = 0.25\,\text{N/cm or } 25\,\text{N/m}$$

Practice

1 A student measures an extension of 24 mm when she hangs a 40 g mass on a spring. Calculate the spring constant in N/m. Show your working.

2 Compare the meaning of the gradient of a graph of force against extension with the meaning of the gradient of a graph of extension against force.

Practice

Exam-style questions

01 A student investigates the extension of a spring.

01.1 Give the name of the pieces of equipment that they could use to measure force and extension. **[2 marks]**

01.2 Describe how they can use this equipment to make the measurements. **[4 marks]**

> **Exam Tip**
> The units for force and extension might give you a clue.

01.3 Explain why the student should take repeat measurements. **[1 mark]**

> **Exam Tip**
> If you only get one result how do you know that it is the correct result?

01.4 Suggest what the student should do if they see a result that does not fit with the pattern of their other results. **[1 mark]**

01.5 Describe the type of graph the student should plot.

Give reasons for your answer. **[2 marks]**

02 A student watches a video about forces.

The video shows a piece of wood floating in a tank of water.

02.1 Write down the name of the non-contact force acting on the wood and the name of the contact force acting on the wood. **[2 marks]**

02.2 The resultant force on the piece of wood is zero.

Describe what that means about the magnitude and direction of the two forces acting on the wood. **[2 marks]**

02.3 In the video the presenter shows that a piece of wood from an ironwood tree will sink rather than float.

The presenter shows this by placing the ironwood on the surface of the water.

It moves down through the water until it reaches the bottom of the tank.

Give the name of **one** other force that is acting on the wood as it moves through the water.

Write down whether it is a contact or a non-contact force. **[2 marks]**

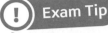

Exam Tip

This question has two parts – make sure you answer both.

---

03 A child is sitting in a supermarket trolley. Her brother is pushing the trolley.

The brother exerts a force of 20 N.

The total distance travelled by the trolley is 30 m.

03.1 Write down the equation which links work done, force, and distance. **[1 mark]**

03.2 Calculate the work done by the brother. **[2 marks]**

03.3 There are **two** units for work done. Name them both. **[1 mark]**

Exam Tip

The equation might give you a clue to one of the units.

03.4 When you lift an object, you do work against gravity.

The brother is doing work when he pushes the trolley at a steady speed.

Name the force against which he is doing work. **[1 mark]**

Exam Tip

What force would cause the trolley to slow down?

03.5 Describe the energy changes when the brother is moving the trolley at a steady speed. **[2 marks]**

04 A student investigates the deflection of a ruler. To do this, they put a ruler between two supports. They tie a piece of string in a loop and hang it in the centre of the ruler. They add weights to the loop, and the centre of the ruler is deflected.

04.1 Suggest a problem the student might find when measuring the deflection of the ruler. **[1 mark]**

04.2 **Table 1** shows the measurements taken by the student.

Table 1

Weight in N	Deflection in mm			Mean deflection in mm
	Repeat 1	Repeat 2	Repeat 3	
0	0	0	0	0
2	3	2	2	2
4	4	5	6	5
6	10	12	17	
8	14	15	15	15
10	17	18	17	17
12	17	18	18	18

Calculate the mean deflection for the weight of 6 N. **[1 mark]**

04.3 Using the axes in **Figure 1**, plot a graph of the data in **Table 1**.
Include a line of best fit. **[4 marks]**

Figure 1

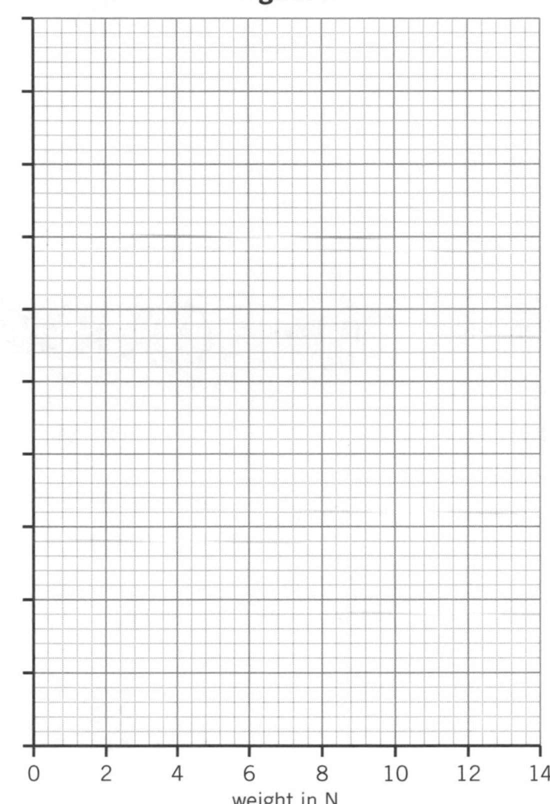

04.4 Determine whether the deflection is proportional to the force.
Explain your answer. **[2 marks]**

> **Exam Tip**
> When ever you're asked to find a mean result be sure to look out for any anomalies!

> **Exam Tip**
> Plot your points using crosses.

> **Exam Tip**
> It's important that you recognise the shape of direct proportion and indirect proportion on a graph as they frequently come up in exam questions.

05 A student is investigating material to make a newtonmeter. They collect data on the stretching of a sample of material. The results are shown in **Table 2**.

Table 2

Weight in g	Length in cm	Length in cm	Length in cm	Average length in cm
100	3.7	3.5	3.5	3.6
200	4.7	4.8	4.1	4.5
300	5.0	5.4	5.2	5.2
400	6.8	7.1	6.8	6.9
500	8.5	9.0	9.2	8.9

05.1 Describe the error the student has made in the first column in **Table 2**. Suggest how to correct it. **[2 marks]**

> **Exam Tip**
> The name of the equipment should give you a clue to the units needed.

05.2 Plot a graph using **Figure 2** of weight against average length. Gravitational field strength = 10 N/kg. **[4 marks]**

> **Exam Tip**
> The question has given you the value for gravitational field strength. This should give you a clue that you need to do a calculation.

Figure 2

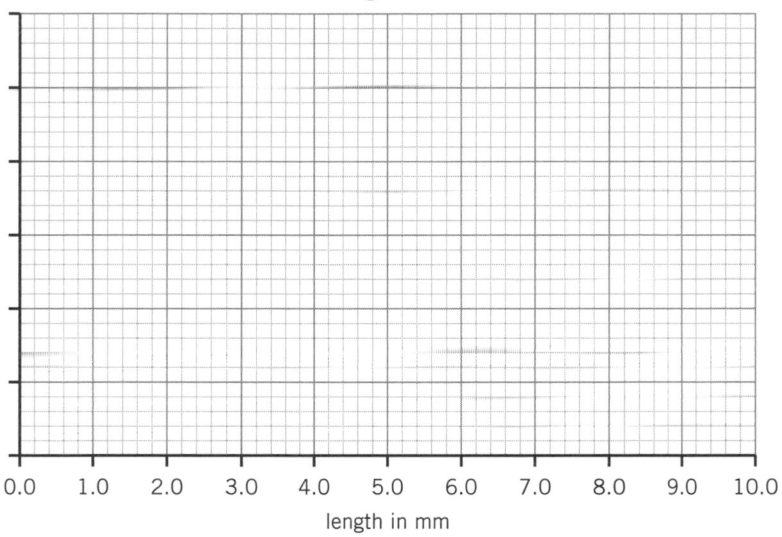

05.3 Estimate the original length of the sample.

Describe your method. **[2 marks]**

> **Exam Tip**
> You'll need to draw on the graph for this!

05.4 Use the shape of the graph to explain how the stiffness of the material changes as the force increases. **[1 mark]**

05.5 The student wants to use the material to make a newtonmeter. Explain why the material would or would not be suitable. **[2 marks]**

06 A student is carrying heavy books in a plastic bag. This has caused the handles of the bag to stretch. Changing the shape of an object by stretching requires two forces. One of these forces is the force of the books on the bag.

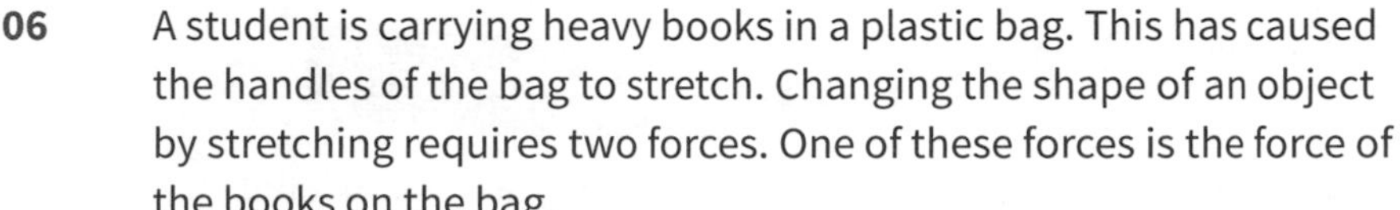

06.1 Name the other force involved in stretching the bag. **[1 mark]**

06.2 When the student takes the book out of the bag there is inelastic deformation of the handles. Define 'inelastic deformation'. **[1 mark]**

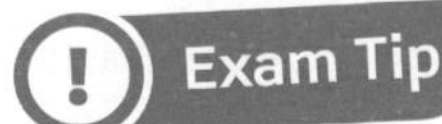

> **Exam Tip**
> Definitions of key words are easy marks in exams – learn them!

06.3 The student cuts the plastic bag into sections then applies different forces to the plastic sections and measures the extension. The student then repeats the experiment with a spring. They plot graphs of their data. One of the graphs is a curved line. The other graph is a straight line. Tick the **two** correct statements. **[2 marks]**

Statement	Correct
The graph for the plastic bag shows a non-linear relationship between force and extension.	
The graph for the plastic bag shows that it is proportional to extension.	
A graph that is a straight line is likely to be for a spring.	
The material that produced a linear graph has been inelastically deformed.	

Exam Tip

If you tick more than two boxes, you won't get full marks even if you pick both the correct answers.

Only tick **two** statements.

07 A student attaches a spring to a retort stand and hangs a weight on the end of the spring. The unstretched length of the spring is 2 cm. The length of the spring when stretched is 3 cm.

07.1 Calculate the extension of the spring in metres. Show how you work out your answer. **[2 marks]**

Exam Tip

The question give you numbers in centimetres, but the answer wants the number in metres. Be careful with your conversions here.

07.2 Write down the equation which links force, extension, and spring constant. **[1 mark]**

07.3 The student used a weight of 2 N. Show that the spring constant of the spring is 200 N/m. **[2 marks]**

Exam Tip

'Show' questions are amazing! You can just keep playing with the numbers until you get the correct answer.

07.4 Calculate the energy stored in the spring. Use the correct equation from the *Physics Equation Sheet*. **[2 marks]**

07.5 The spring is not deformed inelastically. Write down the work done on the spring. Justify your answer. **[2 marks]**

08 A student wants to model how avalanches start. They put a tub with sand in it on an adjustable ramp. **Figure 3** shows the tub of sand when the ramp is flat on the workbench (left) and when it is raised (right).

Figure 3

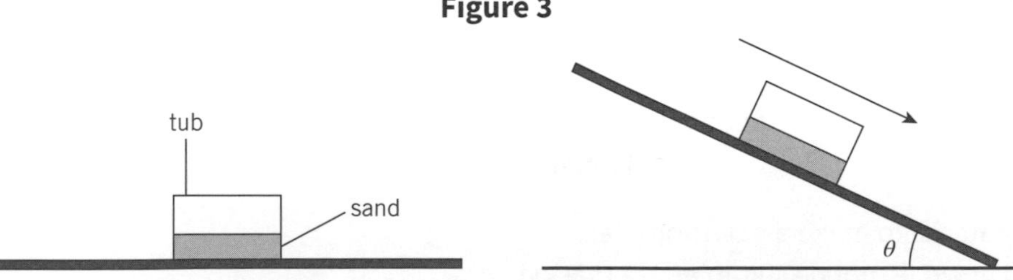

08.1 Draw a free body diagram for the tub of sand when it is flat on the workbench. Label your diagram. **[3 marks]**

Exam Tip

You'll need to label all the arrows.

08.2 The student raises the ramp as shown and writes down the angle (θ) when the tub just starts to move. When the ramp is raised further, the tub accelerates down the ramp. Explain why there is a resultant force on the tub when the ramp is raised. Use ideas about components of forces and friction in your answer. **[3 marks]**

08.3 The student repeats the experiment with different masses of sand in the tub. The results show that the angle of the ramp needed to start the tub moving is independent of the mass of sand. Sketch a graph of angle against mass of sand. **[2 marks]**

Exam Tip

'Sketch' means you just need the shape and axes labels. Numbers and plotted points are not needed.

08.4 Suggest a reason for the shape of the graph. **[1 mark]**

09 A car is travelling on the road. The driving force is 18 000 N. The resistive forces add up to 12 000 N. The weight of the car is 15 kN. The car is travelling from right to left.

09.1 Draw a free body diagram for the car. Label all of the arrows in your diagram. **[5 marks]**

09.2 Calculate the resultant forces in the vertical and horizontal direction. **[3 marks]**

09.3 Calculate the mass of the car. The gravitational field strength is 9.8 N/kg. **[4 marks]**

When ever you need to use gravitational field strength a value will be given to you, you don't need to remember it.

09.4 The average value of the gravitational field strength on Earth is 9.8 N/kg. The gravitational field strength varies between the equator and the poles of the Earth. Suggest how your free body diagram would change if the gravitational field strength was not 9.8 N/kg. **[3 marks]**

10 A dance teacher wants to hang a large mirror in a studio using two lengths of wire. They use two different arrangements as shown in **Figure 4**.

Figure 4

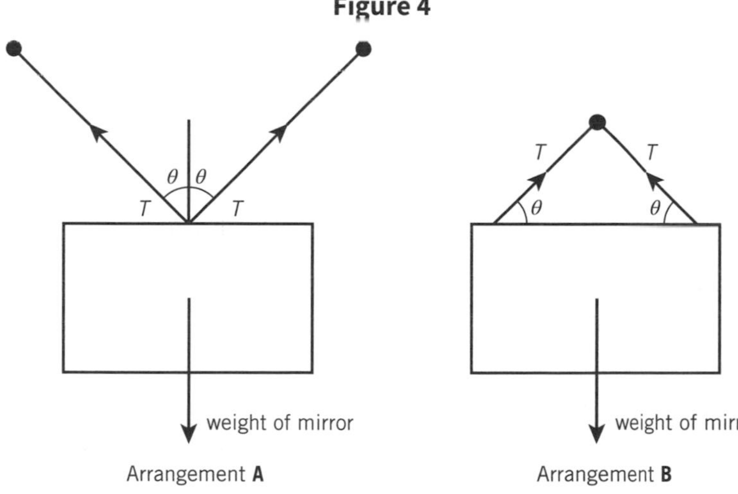

In arrangement **A** the tension in each string = 100 N. The angle $\theta = 40°$. The mirror is hanging and is stationary.

10.1 Use a scale diagram to calculate the weight of the mirror in arrangement **A**. **[3 marks]**

You may be more used to using these skills in maths, but don't be surprised when they turn up in physics.

10.2 Suggest what happens to the tension in the strings when the angle increases in arrangement **A**. Give reasons for your answer. **[2 marks]**

10.3 Another teacher suggests that the force in the string would be less if the mirror was fixed using arrangement **B**. Compare the tensions in the strings in the two arrangements. Assume that the angles are the same. **[4 marks]**

11 A student sees a large crane moving a load. There are counterbalance weights on the other side of a pivot. The load exerts a force of 1000 N. The student estimates that the load is 8 m from the pivot.

11.1 Write down the equation that links the moment of a force, the size of the force, and the distance of the force from the pivot. **[1 mark]**

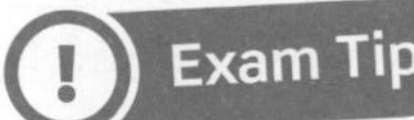

Exam Tip

Don't confuse moment and momentum, they are different things and have different equations.

11.2 Calculate the moment of the force exerted by the load. **[2 marks]**

11.3 The crane pauses. The counterbalance weights are 4 m away from the pivot. The crane is balanced. Calculate the force exerted by the counterbalance weights. Explain how you worked out your answer. **[4 marks]**

Exam Tip

I would alway suggest sketching out a quick diagram to help with this calculation. Add on arrows to show the direction of the forces.

11.4 The crane uses a motor to lift the load. The motor uses gears. Use the words **high** or **low** to complete this sentence. **[1 mark]**

The gears provide a **high / low** speed and a **high / low** turning effect.

12 A student stretched two springs using different forces, and measured the extension. **Figure 5** shows the graph of the results.

Figure 5

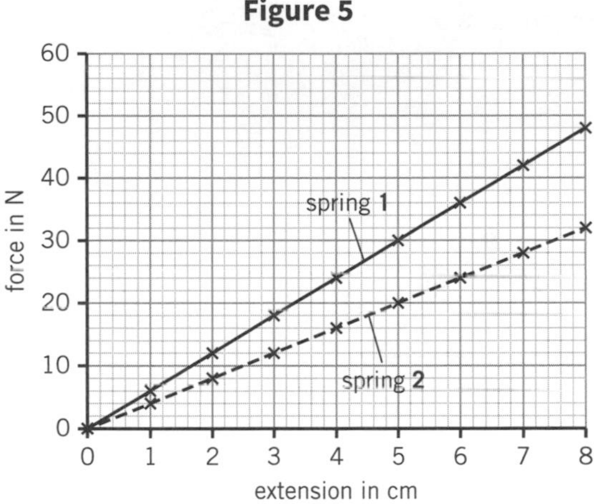

12.1 Show that the force required to stretch Spring **1** by 5 cm is 1.5 times bigger than the force required to stretch Spring **2** by 5 cm. **[2 marks]**

12.2 The student stretched each spring by pulling on both ends. Suggest the difference between the two springs in terms of their stiffness. Explain your answer. **[2 marks]**

Exam Tip

Draw lines on the graph to help you work out which numbers you need to do the calculations with.

12.3 Write the equation that links the spring constant, the extension, and the force applied to a spring **[1 mark]**

12.4 Calculate the spring constant for both springs. **[6 marks]**

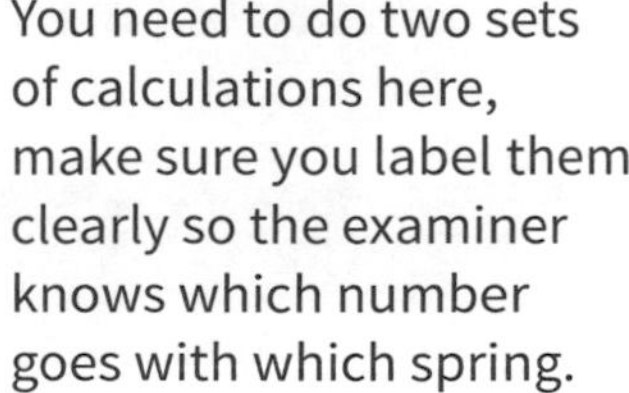

Exam Tip

You need to do two sets of calculations here, make sure you label them clearly so the examiner knows which number goes with which spring.

13 A climber is descending a cliff. For safety, they are attached using a piece of rope. The rope material obeys Hooke's law.

13.1 Write down the equation that links force, the spring constant, and the extension for a material that obeys Hooke's law. **[1 mark]**

13.2 The original length of the rope was 20.00 m. The final length of the rope was 20.14 m. Calculate the extension of the rope. **[1 mark]**

13.3 The climber has a mass of 80 kg. The gravitational field strength is 9.8 N/kg. Calculate the spring constant of the rope. Use an appropriate number of significant figures. **[5 marks]**

Exam Tip

Significant figures is a measure of resolution.

Your answer can't have a higher resolution than the numbers given in the question, so giving too many significant figures would be inappropriate.

13.4 The climber changes the length of the rope to 30 m. A longer rope will have a greater extension. Suggest what happens to the spring constant of the rope.

Explain your answer. **[2 marks]**

14 Scientists attach scientific instruments to robots, which travel to Mars and land on the planet. Here are some data relating to a robot which is sent to Mars:

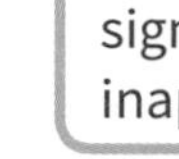
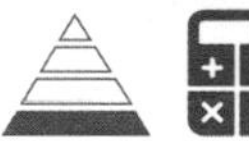

- mass of robot = 140 kg
- gravitational field strength on Earth = 9.8 N/kg

14.1 Write down the equation that links weight, mass, and gravitational field strength. **[1 mark]**

14.2 Calculate the weight of the robot on Earth. **[2 marks]**

14.3 Name the point at which the weight of the robot can be considered to act. **[1 mark]**

14.4 The gravitational field strength on Mars is 3.8 N/kg. Calculate the weight of the robot on Mars and compare this weight with the weight of the robot on Earth. **[5 marks]**

Exam Tip

The value for gravitational field strength is specific to each planet. Be sure you read the question fully to get the right value.

14.5 Identify whether gravitational forces are contact forces or non-contact forces. Justify your answer. **[2 marks]**

Knowledge

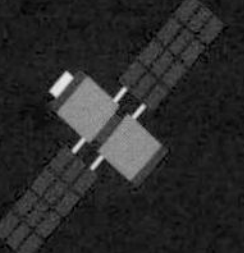   

P10 Pressure in liquids and gases

Fluids

A **fluid** is a substance that can flow.

Fluids are either liquids or gases.

Pressure in a fluid

When the particles of a fluid collide with a surface, such as in a container, they exert a force at right angles (normal) to the surface.

Pressure is the force acting per square metre on a surface.

The unit of pressure is the **pascal** (Pa), which is equal to one newton per square metre.

Pressure can be calculated using:

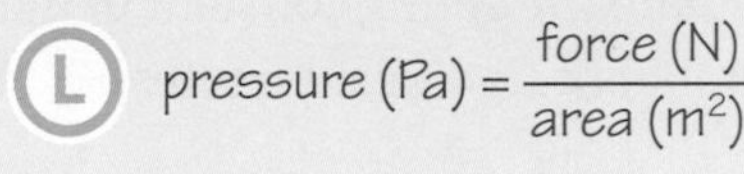

$$\text{pressure (Pa)} = \frac{\text{force (N)}}{\text{area (m}^2\text{)}}$$

$$p = \frac{F}{A}$$

Pressure at depth

The pressure in a liquid increases with the depth of the liquid because:

- the pressure at any point in a liquid is due to the weight of the liquid above that point
- the weight of a liquid depends on its **density**.

Calculating pressure in a column of water

The pressure caused by a column of liquid can be calculated using:

pressure (Pa) = height of the column (m) × density of the liquid (kg/m^3) × gravitational field strength (N/kg)

$$p = h \rho g$$

To calculate the difference in pressure at different depths in a liquid, calculate the pressure at each depth (h) and subtract the smaller value from the larger one.

Revision tip

This equation has both a lower case p (pressure) and a lower case ρ (density) in it – be careful not to get them mixed up!

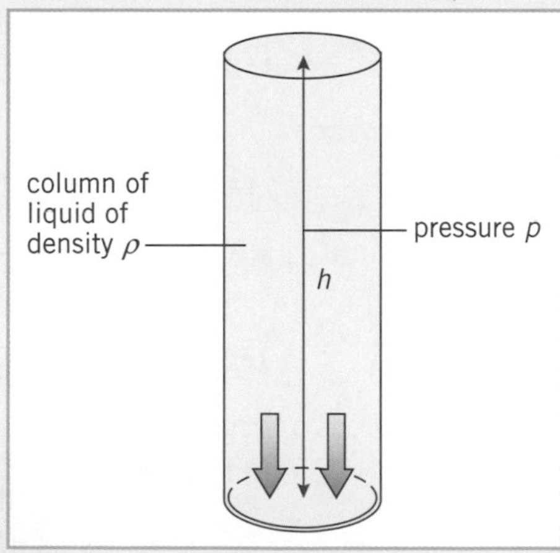

Revision tip

If you were to put holes in the side of this column, the water would spurt out.

Water from a hole in the bottom of the column would spurt out further from the column than a hole at the top, because the hole at the bottom is under more pressure.

Key terms Make sure you can write a definition for these key terms.

altitude atmosphere atmospheric pressure density displace

Upthrust

An object that is partially or completely submerged in a fluid experiences a greater pressure on its bottom surface than its top surface.

This difference in pressure creates an upwards resultant force on the submerged object, known as **upthrust**.

Floating and sinking

An object will sink if its weight is greater than the upthrust.

An object will float if its weight is equal to the upthrust.

Whether an object in a fluid will float or sink depends on its density because:

- the upthrust on an object is equal to the weight of the fluid it **displaces** (pushes out of the way)
- an object that is *more dense* than the fluid will sink because its weight is greater than the weight of the liquid displaced (and so greater than the upthrust)
- an object that is *less dense* than the fluid will float because its weight is less than the weight of the fluid displaced (and so less than the upthrust).

Occasionally an exam question will ask for pressure in non-standard units, for example, N/cm^2.

Don't let this worry you, just use the numbers given in the question and the units they want the answer in as your guide.

Convert to the units they want in the answer – most of the time this is the examiner making the question easier for you.

Atmospheric pressure

The Earth is surrounded by a thin (relative to the size of the Earth) layer of air known as the **atmosphere**.

Air is a fluid, so there is pressure in the atmosphere – this is called **atmospheric pressure**.

Atmospheric pressure is caused by air molecules colliding with surfaces. This decreases as height above a surface (altitude) increases because:

1. there are fewer air molecules in total above the surface as height increases, so the weight of air above the surface decreases
2. density of the atmosphere decreases with altitude, so there are fewer air molecules per cubic metre.

These both mean that atmospheric pressure decreases with increasing altitude because there is less weight of air above the surface.

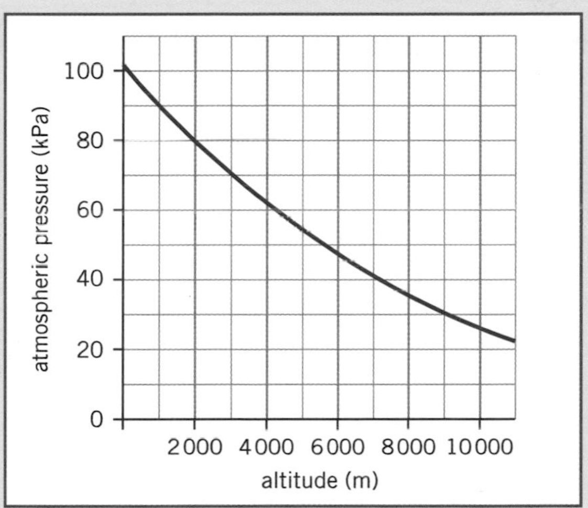

fluid | gravitational field strength | pascal | pressure | upthrust

Retrieval

Learn the answers to the questions below then cover the answers column with a piece of paper and write as many as you can. Check and repeat.

	P10 questions	Answers
1	What is a fluid?	a substance that can flow (liquid or gas)
2	What is the unit of pressure that is equal to one newton per square metre?	pascal (Pa)
3	Why does the pressure in a liquid increase with depth?	pressure at any point in a liquid is due to the weight of the liquid above that point
4	Why does the pressure in a liquid depend on the density of the liquid?	pressure is due to the weight of the liquid, and the weight of a liquid depends on its density
5	What is upthrust?	the resultant force due to the difference in pressure between the top and bottom surfaces of an object submerged in a fluid
6	What will an object placed in a fluid do if its weight is equal to the upthrust?	float
7	What will an object placed in a fluid do if its weight is greater than the upthrust?	sink
8	Why does an object that is more dense than a fluid sink if it is placed in the fluid?	weight of the object is greater than the weight of the fluid displaced, so the weight of the object is greater than the upthrust
9	Why does an object that is less dense than a fluid float if it is placed in the fluid?	weight of the object is less than the weight of the fluid displaced, so the weight of the object is less than the upthrust
10	Does an object that is partially submerged in a fluid experience a greater pressure on its bottom or top surface?	bottom
11	What is the Earth's atmosphere?	the layer of air that surrounds the Earth
12	What is atmospheric pressure caused by?	air molecules colliding with surfaces
13	Why does atmospheric pressure decrease with increased altitude?	the density of the air decreases, fewer air molecules as you go higher – there is less weight of air above a surface and fewer air molecules so density of the atmosphere decreases
14	How does the height of the atmosphere compare to the radius of the Earth?	it is smaller

Put paper here Put paper here Put paper here Put paper here Put paper here Put paper here Put paper here

Now go back and use the questions below to check your knowledge from previous chapters.

Previous questions

Answers

	Previous questions	Answers
1	What is the name of the force acting on an object due to gravity?	weight
2	What is nuclear fusion?	when two light nuclei join to make a heavier one
3	Why is the mass of a substance conserved when it changes state?	the number of particles does not change
4	How does a lever reduce the amount of force needed to create a particular sized moment?	by increasing the distance from the pivot
5	What are the three types of nuclear radiation?	alpha, beta, gamma
6	What is work done?	energy transferred when a force moves an object
7	How small is a nucleus compared to the whole atom?	around 10 000 times smaller
8	Name three greenhouse gases.	water vapour, carbon dioxide, methane
9	What are isotopes?	atoms of the same element (same number of protons) with different numbers of neutrons
10	Name the unit that represents one joule transferred per second.	watt (W)

Put paper here

Required Practical Skills

Area and volume

The unit of **area** is the square metre (m^2).

Since 1 m = 100 cm = 1000 mm, an area of 1 m^2 = 10 000 cm^2 = 1 000 000 mm^2

The area of a rectangle can be calculated using:

area of rectangle = width × length

The area of a triangle can be calculated using:

$$\text{area of triangle} = \frac{(\text{height} \times \text{length of base})}{2}$$

The unit of **volume** is the cubic metre (m^3).

Since 1 m = 100 cm = 1000 mm, a volume of 1 m^3 = 1 000 000 cm^3 = 1 000 000 000 mm^3

The volume of a cuboid can be calculated using:

volume of cuboid = length × width × height

Worked example

A tank of water has the dimensions 2 m width, 4 m height, and 6 m length.

What is the area of the base of the tank?

Answer:

area of rectangle = width × length
= 2 × 6 = 12 m^2

What is the volume of the tank?

Answer:

volume of cuboid = length × width × height
= 6 × 2 × 4 = 48 m^3

Practice

1. A triangle is 2.5 m long and 1.25 m tall. Calculate the area of the triangle.
2. A rectangular container has a base that is 0.30 m by 0.60 m. Calculate the area of the container's base. Give your answer in mm^2.
3. The largest fish tank a pet shop sells is 200 cm long, 60 cm wide, and 90 cm tall. Calculate the volume of the fish tank. Give your answer in m^3.

Practice

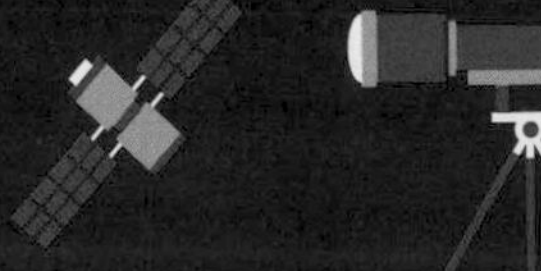

Exam-style questions

01 A gas is held in a container.

01.1 Write the equation that links pressure, force, and area. **[1 mark]**

____

01.2 The force exerted by the gas on each 0.02 m^2 of the container is 2000 N.

Calculate the pressure exerted on the walls of the container. Give the units. **[3 marks]**

Pressure = ______________ Unit = ______________

> **Exam Tip**
>
> Recalling units is worth a whole separate mark. They are easy marks if you take the time to learn them.

01.3 Give the direction of the force exerted on the walls of the container. **[1 mark]**

____

01.4 The gas is replaced by a liquid.

Suggest what happens to the direction of the force. **[1 mark]**

____

____

02 A student is watching a child playing with a plastic boat in a paddling pool.

The plastic boat is partially submerged and floats.

02.1 Identify the forces acting on the plastic boat. **[2 marks]**

____

____

> **Exam Tip**
>
> It can help to sketch an image of the forces with these questions.

02.2 Explain why the plastic boat does not sink. **[2 marks]**

____

____

____

02.3 The child adds sand to the inside of the boat. The boat continues to float.

Explain why the boat still floats. You should include what happens to the forces acting on the boat in your explanation. **[3 marks]**

____

____

____

____

03 Four cubes of the same size are placed in a liquid as shown in **Figure 1**.

Figure 1

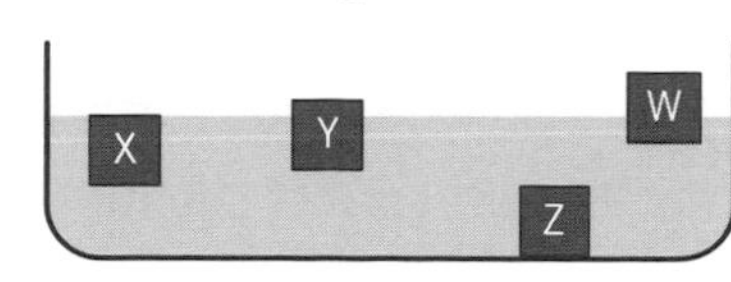

03.1 Objects **W**, **X**, and **Y** will experience different upthrusts. Write down the letters **W**, **X**, and **Y** in order from the largest to the smallest upthrust. **[2 marks]**

03.2 Explain why **Y** is more submerged than **W**. **[3 marks]**

03.3 Write down the equation that links area, force, and pressure. **[1 mark]**

03.4 Cube **X** has a weight of 0.015 N and a side length of 0.01 m. Calculate the average pressure acting on cube **X**. **[3 marks]**

04 A student investigates how a bridge can be lowered and raised using a rope and a pulley.

They draw a model of the equipment they use (**Figure 2**). The rod represents the bridge. The string represents the rope.

Figure 2

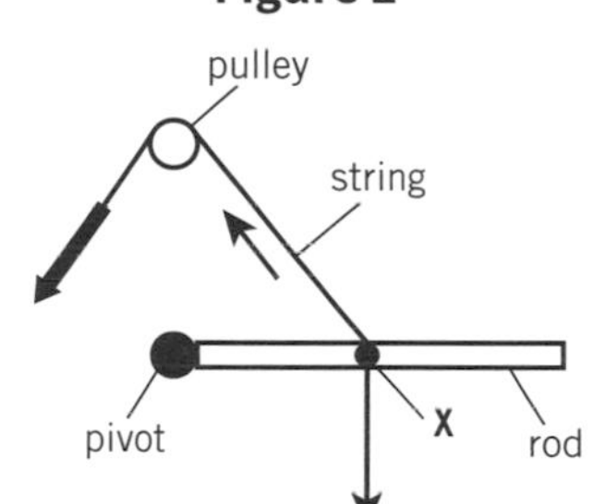

04.1 The weight of the rod acts through point **X**. Give the name of point **X**. **[1 mark]**

04.2 The vertical component of the tension balances the weight of the rod. The tension in the string is 4.0 N. The angle between the string and the rod is 40°. Use a scale diagram to calculate the vertical component of the tension. **[3 marks]**

04.3 Write down the weight of the rod. **[1 mark]**

04.4 The rod slips from the pivot. Only two forces acting on the rod remain – the tension and the weight. Use a scale diagram to find the magnitude of the resultant force acting on the rod. You do not need to find the direction. Assume the tension remains constant. **[3 marks]**

Exam Tip

Drawing simplified diagrams of equipment or experiments is a great way to focus on the important bits of information.

05 A student investigated the link between the pressure of a column of liquid and the height of the column. They used a bottle with markings that showed the volume of water in the bottle. They made a hole near the bottom of the bottle, covered the hole with tape, and then filled the bottle with water.

When the student removed the tape, the water left the bottle. The student measured the distance travelled by the jet of water. They repeated the experiment with different volumes of water in the bottle.

Figure 3

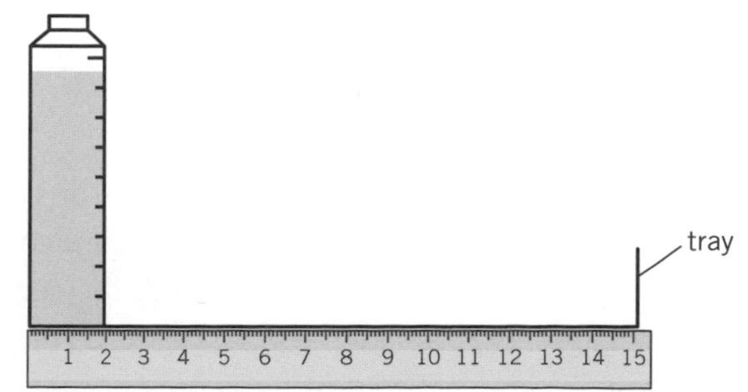

05.1 Give the independent and dependent variable of the investigation. **[2 marks]**

05.2 Sketch the graph that the student will obtain when they plot the data. Explain the shape of the graph. **[4 marks]**

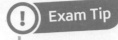

> **Exam Tip**
>
> Sketching a graph means you need to give the overall shape and label the axis but not plot points.

05.3 Describe a source of error due to the set-up of the equipment. Suggest how the student could correct this error. **[2 marks]**

06 Divers use the following rule: water pressure increases by approximately 100 kPa for each 10 m below the water surface.

06.1 Show that the divers' rule is correct. Use the correct equation from the *Physics Equation Sheet*.

Density of water = 1×10^3 kg/m^3
Gravitational field strength = 9.8 N/kg **[3 marks]**

06.2 If the pressure is greater than 405 kPa most divers need to wear a special suit. Calculate the depth at which most divers need to wear this suit. Atmospheric pressure = 101 kPa. **[3 marks]**

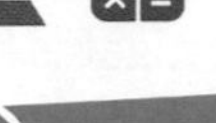

07 A ball is lying on the ground.

07.1 Describe the pairs of interactions that produce the downwards force on the ball and the upwards force on the ball. **[2 marks]**

> **Exam Tip**
>
> It is essential you learn all of the equations from the physics specification, otherwise you won't be able to do the calculations. They're listed at the back of this book for you.

07.2 Write down the equation that links work, force, and distance. **[1 mark]**

07.3 The ball is kicked and it rolls 4.6 m across the grass. The work done against friction is 19 J. Calculate the force of friction acting on the ball. Write your answer to an appropriate number of significant figures. **[4 marks]**

> **Exam Tip**
>
> You might find it easier to put the numbers in before you rearrange the equations.

07.4 The ball is kicked into the air.

Draw a free body diagram of the ball when it is moving horizontally to the right. **[2 marks]**

07.5 When the ball lands on the ground it deforms. Write down the name of the force that does work on the ball to deform it. **[1 mark]**

08 A teacher uses a pressure sensor to show that as he comes down a flight of stairs the atmospheric pressure changes.

08.1 Explain in terms of molecules why the atmosphere exerts a pressure. **[3 marks]**

> **Exam Tip**
>
> When answering **08.1** think about the movement of gases.

08.2 Write down what happens to the atmospheric pressure as the teacher goes down the stairs. **[1 mark]**

08.3 A student says:

'As the teacher goes down the stairs there are fewer air molecules. That is why the pressure changes.'

Do you agree with the student? Explain your answer. **[3 marks]**

09 Scientists use models to explain observations and to make predictions. One atmospheric model suggests that the atmosphere is a layer of gas of uniform density. A student reads that atmospheric pressure halves for every 5 km increase in height above sea level. Use this data to sketch a graph of atmospheric pressure against height for heights between 0 km and 30 km above sea level. Explain the shape of your graph using ideas about the weight of the air. **[4 marks]**

10 Motorbikes have a suspension system. Springs are connected between the wheel and the frame. A motorcyclist of mass 80 kg gets on a motorcycle.

10.1 Write down the equation that links weight, mass, and gravitational field strength.

Units can give you a clue to the rest of the equation. The units of gravitational field strength are N/kg. This is saying that to calculate gravitational field strength, you need to divide Newtons by kilograms. This is a big hint to the equation.

10.2 Calculate the weight of the motorcyclist. Gravitational field strength = 9.8 N/kg. **[2 marks]**

10.3 There are four springs on the motorbike. The motorcyclist's weight is supported by all four springs. Calculate the force applied to each spring. Show your working. **[2 marks]**

10.4 When the motorcyclist sits on the bike, each spring compresses a distance of 3.4 cm. Use the equation that links force, extension, and spring constant to calculate the spring constant of the spring. **[4 marks]**

Remember there are four springs!

10.5 When the motorcyclist gets off the bike the springs return to their original length. Suggest whether the work done on the spring is bigger than, smaller than, or equal to the stored elastic potential energy. Explain your answer. **[2 marks]**

11 A diver is testing his water pressure gauge in the ocean. He measures the pressure of the water at a depth of 1 m.

11.1 Calculate the pressure due to a column of water that is 1 m high. Use the correct equation from the *Physics Equation Sheet*. The density of water = 10^3 kg/m^3. Gravitational field strength = 9.8 N/kg. **[2 marks]**

11.2 The actual reading on the pressure gauge is much bigger than the number calculated in **06.1**. Explain why the gauge reads a number that is much bigger. **[2 marks]**

Even if you don't have an answer for **11.1** you can still answer this question.

11.3 Salt water has a higher density than pure water. The density of the sea water can change depending on the salt content. Suggest what the diver would see if he looked at the pressure gauge as he moved through areas with different salt content at the same depth. Explain your answer. **[3 marks]**

12 A mountaineer has a small sealed bag full of air at the bottom of the mountain. He notices that the volume of the bag has increased by a factor of three once he reaches the top of the mountain.

12.1 Describe how the pressure of the air changes as the mountaineer goes up the mountain. **[3 marks]**

12.2 Calculate the pressure at the top of the mountain. Atmospheric pressure at the bottom of the mountain = 100 kPa. **[2 marks]**

12.3 Suppose the density of the air is not constant but decreases with height. Suggest what effect this will have on your answer to **12.1**. **[1 mark]**

Knowledge

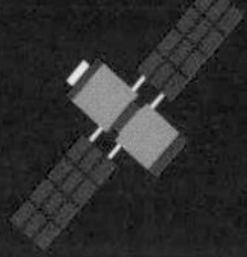
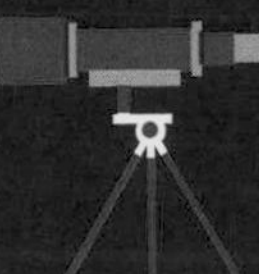

P11 Speed

Distance and displacement

Distance:

- is how far an object moves
- is a scalar quantity so does not have direction.

Displacement is a vector and includes the *distance* and *direction* of a straight line from an object's starting point to its finish point.

Velocity

The **velocity** of an object is its speed in a given direction.

Velocity is a vector quantity because it has a magnitude and direction.

An object's velocity changes if its direction changes, even if its speed is constant.

An object moving in a circle can have a constant speed but its velocity is always changing because its direction is always changing.

Acceleration

Acceleration is the change in velocity of an object per second. It is a vector quantity.

The unit of acceleration is metres per second squared, m/s^2.

An object is accelerating if its speed or its direction (or both) are changing. A negative acceleration means an object is slowing down, and is called **deceleration**.

Acceleration can be calculated using:

$$\text{acceleration } (m/s^2) = \frac{\text{change in velocity } (m/s)}{\text{time taken } (s)}$$

$$a = \frac{\Delta v}{t}$$

Uniform acceleration is when the acceleration of an object is constant.

The following equation applies to objects with uniform acceleration:

$$(\text{final velocity})^2 - (\text{initial velocity})^2 = 2 \times \text{acceleration} \times \text{distance}$$

$$v^2 - u^2 = 2as$$

Speed

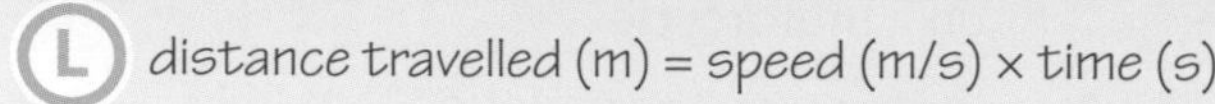

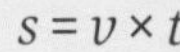

The symbol for distance is s, and the symbol for speed is v.

In reality, objects rarely move at a constant speed. So it can be useful to calculate average speed:

$$\text{average speed } (m/s) = \frac{\text{total distance travelled } (m)}{\text{total time taken } (s)}$$

Some typical average speeds are:

- walking ≈ 1.5 m/s
- running ≈ 3 m/s
- cycling ≈ 6 m/s

The speed of sound and the speed of the wind also change depending on the conditions. A typical value for the speed of sound is 300 m/s

Revision tip

Try going over all the formulas we use in physics and seeing which quantities are vector and which are scalar.

Drag forces

When an object moves through a fluid (liquid or gas) a frictional force drags on it.

These drag forces:

- always act in the opposite direction to an object's movement
- increase with the object's speed – the greater the speed, the greater the frictional force
- depend on the shape and size of the object.

Streamlining an object reduces the drag it experiences.

The frictional drag force in air is called **air resistance**.

Key terms Make sure you can write a definition for these key terms.

air resistance deceleration displacement streamlining terminal velocity uniform acceleration velocity

Distance–time graphs

A distance–time graph shows how the distance travelled by an object travelling in a straight line changes with time.

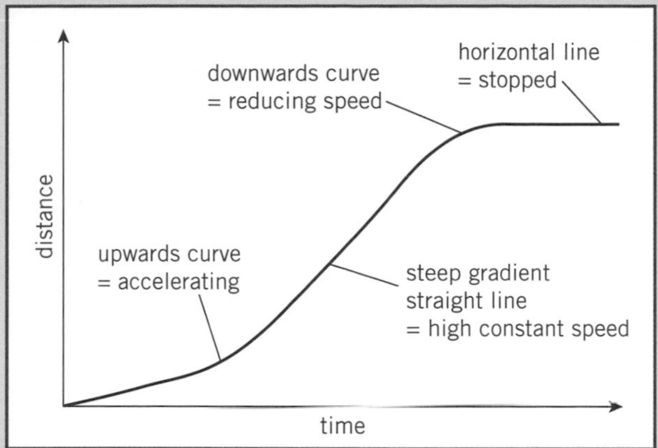

The gradient of the line in a distance–time graph is equal to the object's speed.

If the object is accelerating, the speed at any time can be found by calculating the gradient of a tangent to the curved line at that time.

Velocity–time graphs

A velocity–time graph shows how the velocity of an object changes with time.

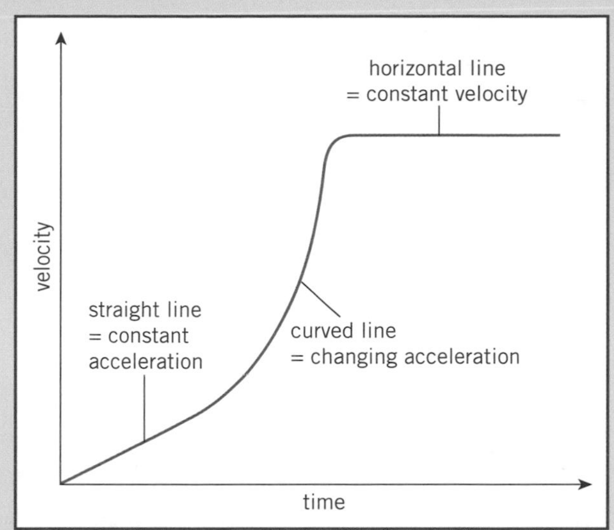

The gradient of the line in a velocity–time graph is equal to the object's acceleration.

The area under the line on a velocity–time graph represents the distance travelled (or displacement).

Terminal velocity

For an object falling through a fluid:

- there are two forces acting – its weight due to gravity and the drag force
- the weight remains constant
- the drag force is small at the beginning, but gets bigger as it speeds up
- the resultant force will get smaller as the drag force increases
- the acceleration will decrease as it falls
- if it falls for a long enough time, the object will reach a final steady speed.

Terminal velocity is the constant velocity a falling object reaches when the frictional force acting on it is equal to its weight.

Graph of terminal velocity

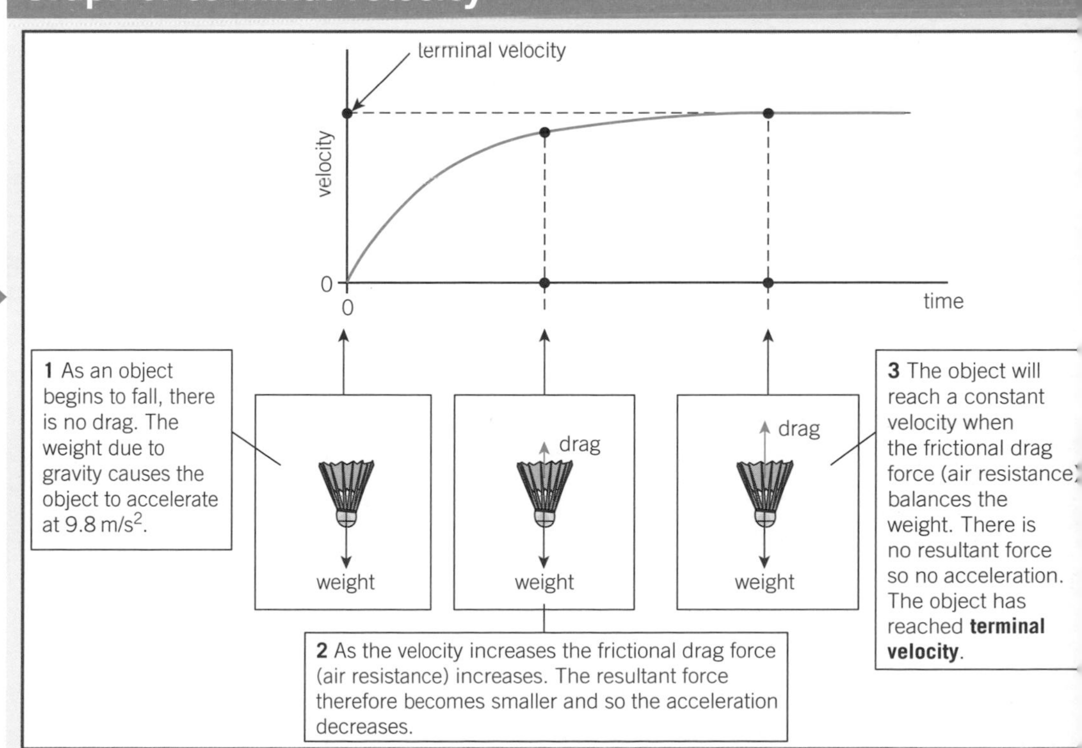

1 As an object begins to fall, there is no drag. The weight due to gravity causes the object to accelerate at 9.8 m/s^2.

2 As the velocity increases the frictional drag force (air resistance) increases. The resultant force therefore becomes smaller and so the acceleration decreases.

3 The object will reach a constant velocity when the frictional drag force (air resistance) balances the weight. There is no resultant force so no acceleration. The object has reached **terminal velocity**.

Retrieval

Learn the answers to the questions below then cover the answers column with a piece of paper and write as many as you can. Check and repeat.

	P11 questions	Answers
1	What is the difference between distance and displacement?	distance is a scalar quantity and only has a magnitude (size), displacement is a vector quantity and has both magnitude and direction
2	What is the difference between speed and velocity?	speed is a scalar quantity and only has a magnitude (size), velocity is a vector quantity and has both magnitude and direction
3	What factors can affect the speed at which someone walks, runs, or cycles?	age, fitness, terrain, and distance travelled
4	What are typical speeds for a person walking, running, and cycling?	1.5 m/s, 3.0 m/s, and 6.0 m/s respectively
5	What are typical speeds of a car and a train?	13–30 m/s and 50 m/s respectively
6	What is a typical speed for sound travelling in air?	330 m/s
7	What is acceleration?	change in velocity of an object per second
8	What is the unit of acceleration?	m/s^2
9	How can an object be accelerating even if it is travelling at a steady speed?	if it is changing direction
10	What is happening to an object if it has a negative acceleration?	it is slowing down
11	What information does the gradient of the line in a distance–time graph provide?	speed
12	What information does the gradient of the line in a velocity–time graph provide?	acceleration
13	How can the distance travelled by an object be found from its velocity–time graph?	calculate the area under the graph
14	What is the acceleration of a free-falling object due to gravity?	~9.8 m/s^2
15	What is the name for the steady speed a falling object reaches when the resistive force is equal to its weight?	terminal velocity
16	What is the general name for the frictional forces an object experiences when moving through a fluid (liquid or gas)?	drag
17	In which direction does the drag on an object always act?	in the direction opposite to which it is moving
18	What happens to the drag on an object as its speed increases?	the drag increases
19	What can be done to reduce the drag on an object?	streamlining

Put paper here

Now go back and use the questions below to check your knowledge from previous chapters.

Previous questions | Answers

	Previous questions	Answers
1	What is a fluid?	a substance that can flow (liquid or gas)
2	What is upthrust?	the resultant force from the difference in pressure between the top and bottom surfaces of an object submerged in a fluid
3	How can an electron move up an energy level?	absorb sufficient electromagnetic radiation
4	What does a material's thermal conductivity tell you?	how well it conducts heat
5	Which two quantities do you need to measure to find the density of a solid or liquid?	mass and volume

Put paper here

Put paper here

Maths Skills

Practise your maths skills using the worked example and practice questions below.

Area of graphs

The area between the line of a graph and the x axis sometimes represents a useful quantity.

To find the area under a graph made up of straight lines, split the shape under the line into rectangles and triangles and add their individual values together.

If the graph has curved sections, estimate the area under the curve by following these steps:

1. calculate the area of a single small square on the graph
2. count the whole and half squares under the curved part of the line
3. multiply the number of squares by the area per square.

Worked example

The graph shows how the velocity of an object changes with time. How far does the object travel during 24 seconds?

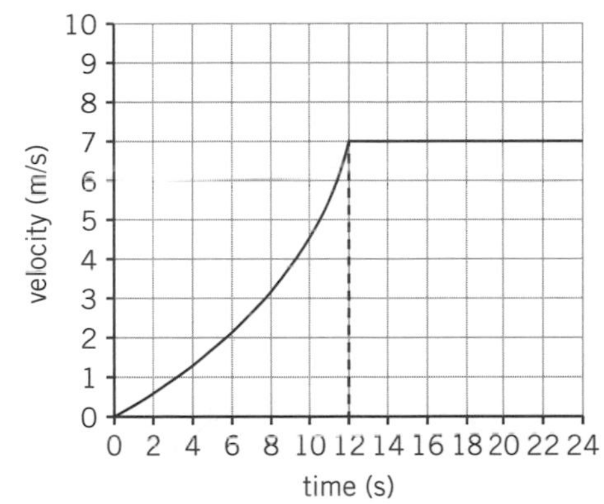

Calculate the area of the rectangular section:

$$7\,\text{m/s} \times (24 - 12)\,\text{s} = 7 \times 12 = 84\,\text{m}$$

Calculate the area a single square represents:

$$1\,\text{m/s} \times 2\,\text{s} = 2\,\text{m}$$

Count the number of whole squares under the curved part of the line (adding any half squares together) = 16

Multiply the number of squares under the curved line by the area one square represents:

$$16 \times 2 = 32\,\text{m}$$

Add the area under the curved section to the area of the rectangular section:

$$84 + 32 = 116\,\text{m}$$

Practice

The graph below shows a velocity time graph for a cyclist:

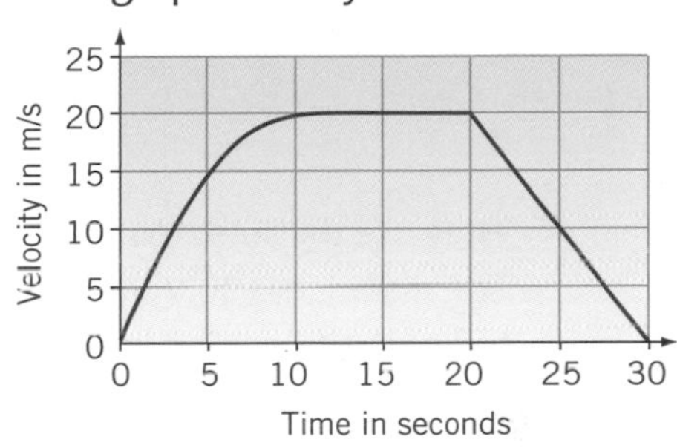

1. Describe the cyclist's motion in the time between 10 and 20 seconds.
2. Calculate the acceleration of the cyclist at 8 seconds.
3. Calculate the total distance travelled by the cyclist in the first 20 seconds.

Exam-style questions

01 A student sets up some equipment to measure the acceleration due to gravity as shown in **Figure 1**.

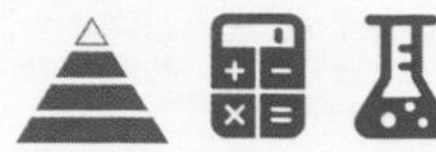

Figure 1

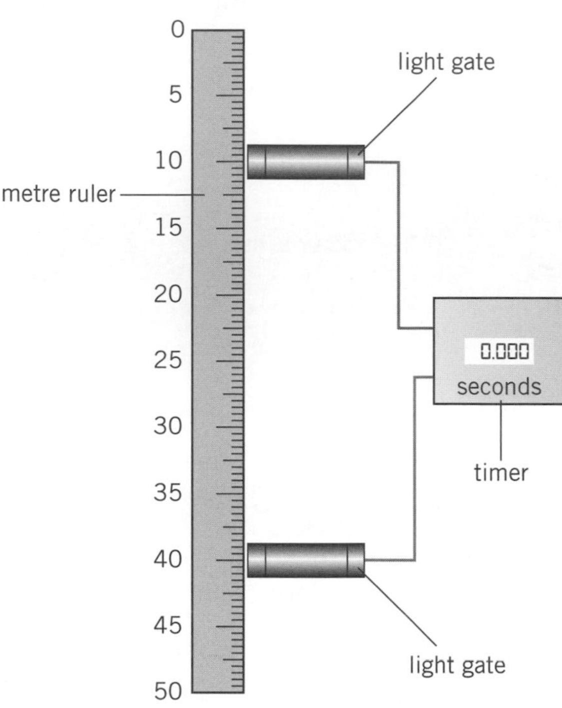

They use the ruler to measure the distance between the light gates. The distance is 30.0 cm. The student drops a piece of card from the top of the ruler. The light gates are connected to a data logger.

The student records the measurements in **Table 1**.

Exam Tip

Look out for non-standard units. Here, the distance is measured in cm, not m.

Table 1

Attempt	Velocity at light gate 1 in m/s	Velocity at light gate 2 in m/s
1	1.376	2.987
2	0.254	2.664
3	1.769	3.187

01.1 Describe how she should use the ruler to get an accurate measurement of the distance between the light gates. **[2 marks]**

______

______

______

01.2 Suggest why the velocities of the card at light gate 1 are different. **[1 mark]**

______

______

Exam Tip

Look at the equipment used to record each bit of data.

01.3 Explain why there are more significant figures in the velocities recorded in the table than in the distance measurements. **[3 marks]**

____

____

____

____

____

01.4 Calculate the acceleration due to gravity using the measurements for attempt 1. Use the correct equation from the *Physics Equations Sheet*.

Give your answer to an appropriate number of significant figures. **[3 marks]**

acceleration = ______________ m/s^2

01.5 Suggest **one** reason why the measured value of the acceleration due to gravity is different from the agreed value. **[1 mark]**

____

____

02 Two students are travelling to school. Student **A** is walking. Student **B** is cycling.

02.1 Write down the typical speeds for students **A** and **B**. **[2 marks]**

____

____

Exam Tip

These estimated values come straight from the specification.

02.2 Write down the equation that links distance travelled, speed, and time. **[1 mark]**

____

02.3 Both students travel 1.5 km.

They leave their houses at the same time.

Calculate the difference in the times that they take to reach the school. Give your answer in minutes. **[6 marks]**

Exam Tip

You need to carry out two different calculations, so make sure you lay out your working clearly. If the examiner can't read your answer, they can't give you the marks for workings.

difference in time = ______________ minutes

02.4 Student **B** monitors his speed. He sees that his fastest speed is much faster than his average speed.

Explain why. **[2 marks]**

____

____

____

03 **Table 2** shows the data for a student in a race on sports day.

Table 2

Time in s	Distance in m
0	0
2	2
4	5
6	8
8	14
10	20
12	22

Figure 2

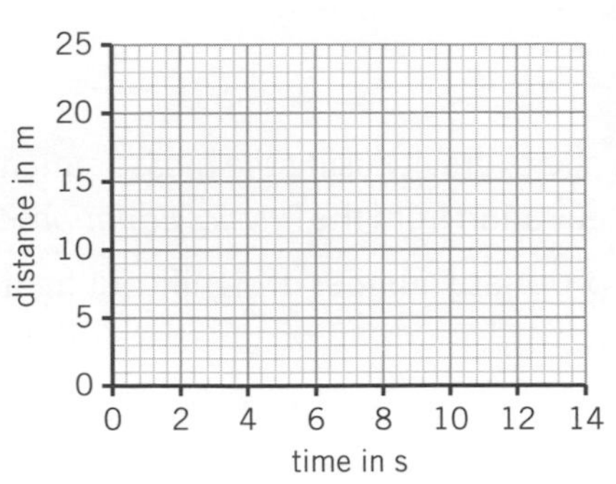

Exam Tip

Always plot points using crosses and use a pencil in case you go wrong.

03.1 Plot the data on **Figure 2**. **[2 marks]**

03.2 Calculate the speed of the student at a time of 4 seconds. **[3 marks]**

03.3 Describe the motion of the student between 6 seconds and 12 seconds. **[2 marks]**

Exam Tip

You should show your working on the graph.

04 A group of students collect data for a car journey. The journeys are shown as lines on the graph in **Figure 3**.

Figure 3

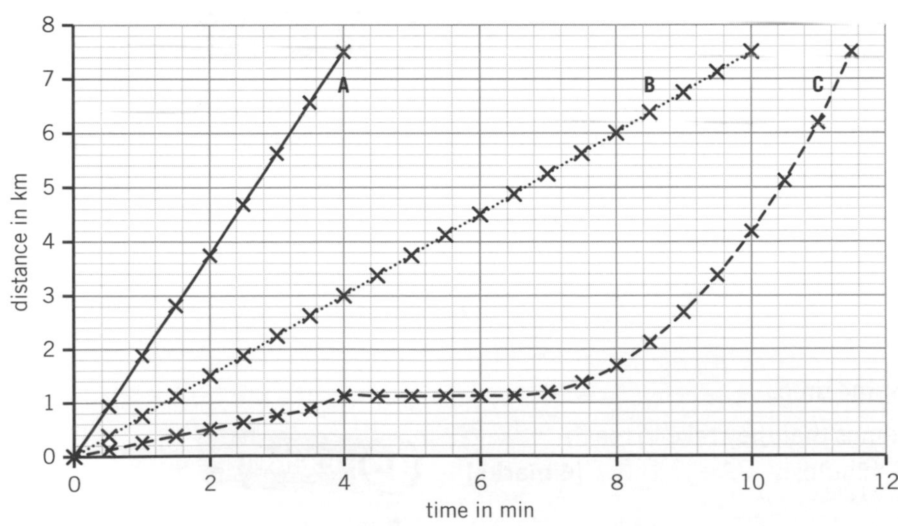

Exam Tip

You might find it helpful to annotate the graph with notes on when the cars are not moving, moving at a steady speed, or accelerating.

04.1 Which car travelled at the highest steady speed? Give reasons for your answer. **[2 marks]**

04.2 Which car or cars stopped at some point during the journey? Give reasons for your answer. **[2 marks]**

Exam Tip

Use data from the graph to support your answer.

04.3 Which car or cars accelerated? Give reasons for your answer. **[2 marks]**

04.4 All three cars travelled the same overall distance. Which car has the highest average speed? Give reasons for your answer. **[2 marks]**

04.5 Use the graph to calculate the highest average speed. **[4 marks]**

05 A teacher shows the class a spinning paper helicopter. They drop the helicopter and use a datalogger to record how the velocity changes over time. **Figure 4** shows the velocity–time graph for the helicopter.

Figure 4

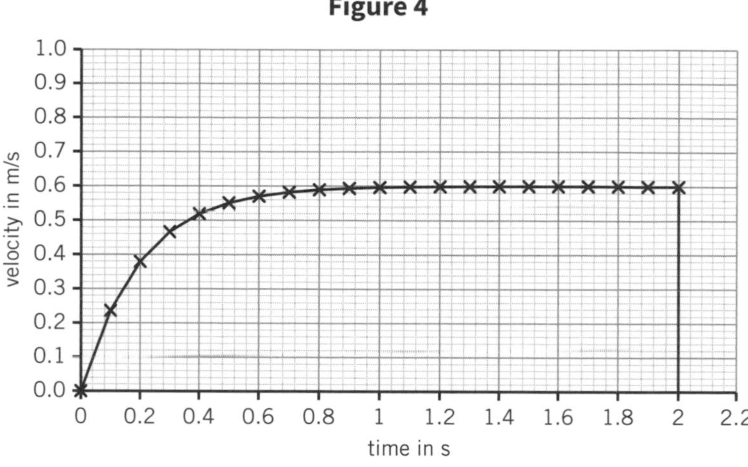

05.1 Write down the terminal velocity of the helicopter. **[1 mark]**

05.2 Use **Figure 4** to estimate the distance travelled by the helicopter. Describe your method. **[3 marks]**

05.3 The teacher adds paper clips to the body of the helicopter and drops the helicopter again from the same height. On **Figure 4**, sketch the graph for the helicopter with additional paper clips. Assume that the helicopter reaches terminal velocity. Explain the shape of the line that you have drawn. **[6 marks]**

05.4 Suggest **one** reason why the acceleration of the helicopter is not constant. **[1 mark]**

06 A person needs to travel 20 miles to work. He has a choice of three different methods of transport to use – car, bicycle, and train.

Compare the time it would take, on average, to travel 20 miles to work by these methods. Choose suitable units in which to do your calculations. There are 1609 m in 1 mile. Write down any assumptions that you make. Suggest the effect on your answers if you do not make these assumptions. **[6 marks]**

07 Many drivers use satellite navigation (satnav) devices when driving to a destination. The satnav communicates with a global positioning system (GPS) satellite. A GPS satellite is at a distance of 20 200 km above the surface of the Earth. The signals from the satnav device to the satellite travel at a speed of 3×10^8 m/s.

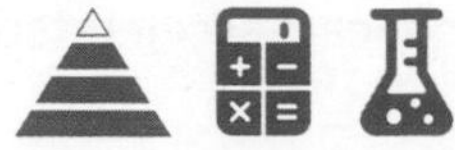

07.1 Show that the time it takes the signal to travel from the car to the satellite and back is about 0.1 s. **[3 marks]**

Exam Tip

In a 'show' question, you've already been given the answer. If you're not exactly sure what to do, work backwards from the numbers given, using the correct equations. Once you've got it, make sure you make it clear what your final answer is to the examiner.

07.2 A car is travelling at 55 mph. Calculate the distance the car would have moved in the time it takes the signal to travel to the satellite and back. There are 1609 m in 1 mile. There are 3600 s in 1 hour. **[3 marks]**

07.3 To pinpoint the position of the car, the satnav device communicates with at least three satellites. This reduces the error in the position to about 30 cm. Suggest whether the error is a systematic or a random error. Explain your answer. **[2 marks]**

07.4 Suggest how the satnav device calculates the speed of the car. **[3 marks]**

07.5 The GPS satellite is orbiting the Earth at a constant speed. Explain how it can be accelerating at the same time as it is moving with a constant speed. **[2 marks]**

08 A block of stone has a mass of about 10 tonnes. One tonne = 10^3 kg. The block has to be moved a vertical distance of 2 m.

08.1 Write down the name of the force that work is done against to move the block vertically 2 m. **[1 mark]**

08.2 Calculate the weight of the block. Gravitational field strength = 9.8 N/kg. **[3 marks]**

08.3 Calculate the work done lifting the block a vertical distance of 2 m. **[3 marks]**

08.4 The block is pulled up a 4 m long ramp. The top of the ramp is 2 m above the ground. Describe how a student can use a scale diagram to calculate the magnitude of the component of the weight parallel to the slope. **[3 marks]**

08.5 The student works out that the component of the weight parallel to the slope is 50 000 N. As the block is being pulled, the force acting on the block due to friction is 3000 N. Calculate the work done pulling the stone up the ramp using the resultant of these two forces. **[3 marks]**

08.6 The work done using the ramp is larger than the work done lifting the stone directly. Suggest why a ramp is used. **[1 mark]**

09 A student collects data by stretching a spring and measuring the extension. They plot the graph shown in **Figure 5**.

Figure 5

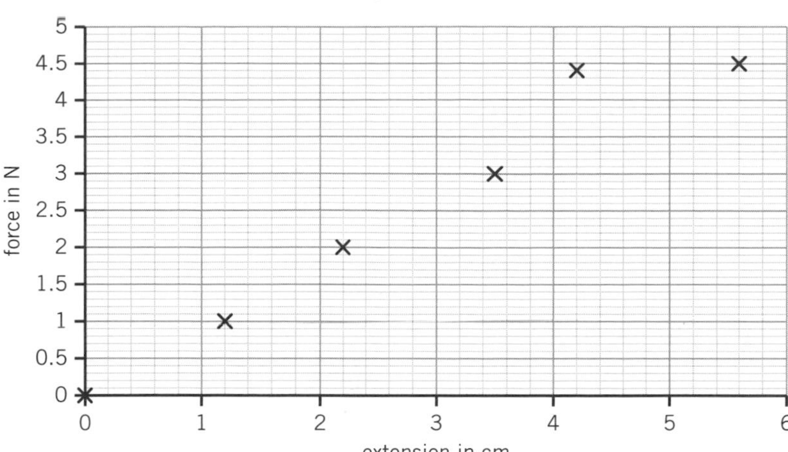

09.1 Suggest which reading is an outlier. **[1 mark]**

09.2 Write the equation that links force, spring constant, and extension. **[1 mark]**

09.3 Use the graph to find the spring constant of the spring. **[4 marks]**

09.4 Describe the shape of the graph if the limit of proportionality is exceeded. **[1 mark]**

Exam Tip

You need to draw a line of best fit first. Make sure you do not include the outlier from **09.1** when drawing it.

10 A student puts a trolley at the top of a ramp. Two light gates are placed on the ramp so that the trolley can move through them. One light gate is placed close to the top of the ramp. The other is placed at the bottom of the ramp. Each light gate measures the velocity of the trolley as it passes:

- initial velocity = 1.12 m/s
- final velocity = 7.12 m/s
- time taken to travel between the light gates = 1.25 s

Exam Tip

This is required practical, but there are lots of different ways this can be done. You can get the same data without light gates if your school doesn't have them.

10.1 Write down the equation that links acceleration, velocity, and time. **[1 mark]**

10.2 Calculate the acceleration of the trolley on the ramp. Give the correct unit with your answer. **[3 marks]**

10.3 Compare this acceleration with the acceleration due to gravity. Give your answer as a ratio. **[2 marks]**

10.4 The trolley leaves the ramp with the final velocity and moves on a rough section of the floor. The trolley takes 5 seconds to stop. The data logger plots a velocity–time graph for the trolley when it is both accelerating and decelerating. Sketch the graph the datalogger makes. Assume that the acceleration and deceleration are constant. Explain the shape of the graph. **[5 marks]**

Exam Tip

A sketched graph just shows the overall shape of the line. You do not need to plot any points or add any values.

11 A scientist uses a sensor to measure the pressure difference between the surface and the bottom of a lake.

11.1 Explain how and why the pressure changes between the top and bottom of the lake. **[2 marks]**

11.2 The scientist makes their measurements. Pressure at the surface of the lake = 101 325 Pa. Pressure at the bottom of the lake = 1318250 Pa. Use the correct equation from the *Physics Equations Sheet* to calculate the change in height that produced this change in pressure. Gravitational field strength = 9.8 N/kg. Density of water = 1000 kg/m^3. **[4 marks]**

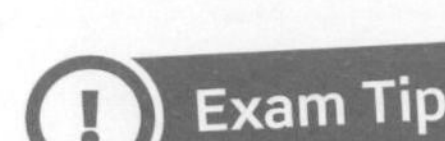

Use the ideas of particles moving to help answer this question.

12 A model rocket contains two fuel cells. When the fuels burn, they force air from the rocket. This produces a force upwards on the rocket. **Figure 6** shows the first 13 seconds of the rocket's journey. Section **AB** shows when the first fuel cell burns to accelerate the rocket upwards. Section **BC** shows when the second fuel cell burns to accelerate the rocket again. When the second fuel cell has finished the only force acting on the rocket is the weight of the rocket.

Figure 6

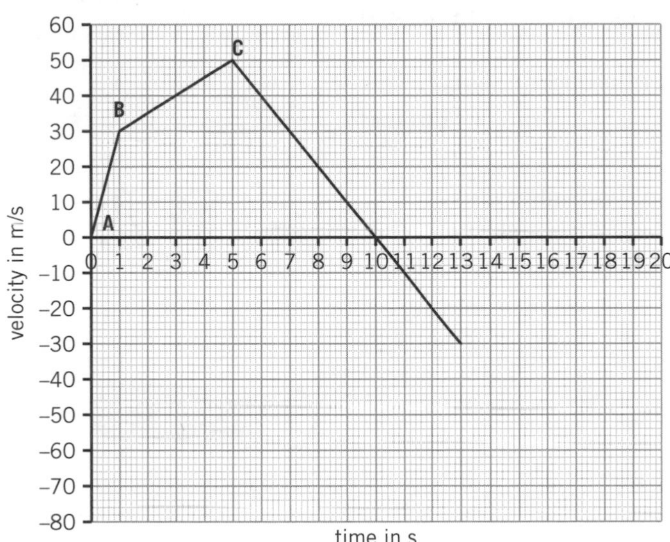

Exam Tip

Go through the question with some highlighters and mark the corresponding part of the graph with the same colour.

12.1 Suggest which of the fuel cells burns faster. Explain your answer in terms of energy. **[2 marks]**

12.2 Use **Figure 6** to calculate the acceleration of the rocket during section **BC**. **[2 marks]**

Exam Tip

You'll have to draw lines on the graph to help with your working.

12.3 Use **Figure 6** to write down the time at which the rocket reaches the maximum distance from the ground. Give reasons for your answer. **[2 marks]**

12.4 The equipment recording the velocity of the rocket stopped recording after 13 seconds. Explain how you know that the rocket had not hit the ground at that time. **[2 marks]**

13 A student investigates terminal velocity by dropping balls of modelling clay in a liquid. She measures the velocity just before the ball enters the liquid and again after a certain distance.

13.1 Identify suitable equipment that she can use to measure the velocity of the ball. **[1 mark]**

13.2 For a particular ball, she makes the following measurements:

- initial velocity = 3.5 cm/s
- final velocity = 1.3 cm/s
- distance over which the velocity changes = 30 cm

Use the correct equation from the *Physics Equations Sheet* to calculate the average acceleration of the object. Justify the sign of the acceleration. **[4 marks]**

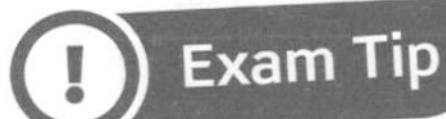

Acceleration can be positive or negative. A negative acceleration is often called deceleration.

13.3 The student wonders if the modelling clay reaches terminal velocity. Suggest how the student could take measurements to check whether terminal velocity has been reached. **[2 marks]**

13.4 The student repeats the experiment with fluids of different densities. She measures the distance over which the velocity changes from 3.5 m/s to 1.3 m/s.

Suggest how and why the distance depends on the density. **[3 marks]**

14 A student has learned about scalars and vectors.

14.1 They know that two of the quantities they learnt about are vectors, and two are scalars.

Choose **one** answer. **[1 mark]**

speed and velocity are vectors

displacement and velocity are scalars

distance and speed are vectors

distance and speed are scalars

14.2 A cyclist starts from home and travels 10 km north, then 20 km south. Calculate the final displacement of the cyclist from home. **[2 marks]**

14.3 Explain why the total distance travelled is not the same as the final displacement. **[2 marks]**

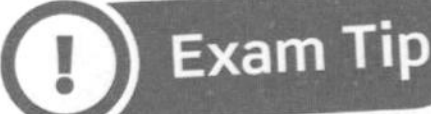

For this question you will need to be clear on the difference between speed and velocity.

14.4 The cyclist says:

'I travelled at 6 km/h north, then 10 km/h south.'

Is the cyclist describing their speed or velocity? Give reasons for your answer. **[2 marks]**

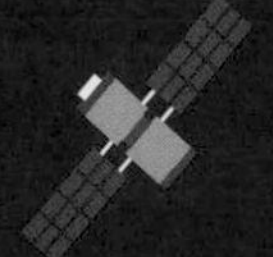

Knowledge

P12 Newton's laws of motion

Newton's First Law

The **inertia** of an object is its tendency to remain in a steady state, i.e., at rest or moving in a straight line at a constant speed.

Newton's First Law says that the velocity, speed, and/or direction of an object *will only change if a resultant force is acting on it*.

This means that:

- If the resultant force on a stationary object is zero, the object will remain stationary.
- If the resultant force on a moving object is zero, it will continue moving at the same velocity, in a straight line.
- If the resultant force on an object is not zero, its velocity *will* change.

When a car is travelling at a steady speed, the resistive forces (e.g., friction and air resistance) must be balanced with the driving forces.

A change in velocity can mean an object:

- starts to move
- stops moving
- speeds up
- slows down
- changes direction.

There *must* be a resultant force acting on an object if it is doing *any* of the things listed above.

Momentum

Momentum is a property of all moving objects. It is a vector quantity.

Momentum depends on the mass and velocity of an object and is defined by the equation:

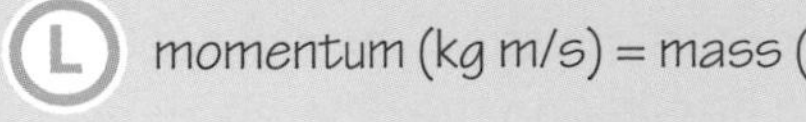

momentum (kg m/s) = mass (kg) × velocity (m/s)

$$p = mv$$

Newton's Second Law

Newton's Second Law says that the acceleration a of an object:

- is proportional to the resultant force on the object

$$a \propto F$$

- is inversely proportional to the mass of the object

$$a \propto \frac{1}{m}$$

Resultant force, mass and acceleration are linked by the equation:

resultant force (N) = mass (kg) × acceleration (m/s²)

$$F = ma$$

The **inertial mass** of an object is a measure of how difficult it is to change the velocity of an object. It can be found using:

$$\text{inertial mass (kg)} = \frac{\text{force (N)}}{\text{acceleration (m/s}^2\text{)}}$$

$$m = \frac{F}{a}$$

Law of Conversion Momentum

The **Law of Conservation of Momentum** says that:

In a closed system, the total momentum before an event (a collision or an explosion) is *equal* to the total momentum after the event.

If two moving objects collide the law of conservation can be written as:

$$m_1u_1 + m_2u_2 = m_1v_1 + m_2v_2$$

m_1 = mass of object 1 m_2 = mass of object 2

u_1 = initial velocity of object 1 u_2 = initial velocity of object 2

v_1 = final velocity of object 1 v_2 = final velocity of object 2

Key terms Make sure you can write a definition for these key terms.

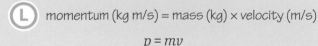

force pair inertia inertial mass momentum recoil

Newton's Third Law

Newton's Third Law states that whenever two objects interact with each other, they exert *equal and opposite* forces on each other.

This means that forces always occur in pairs.

Each pair of forces:

- act on separate objects
- are the same size as each other
- act in opposite directions along the same line
- are of the same type, for example, two gravitational forces or two electrostatic forces.

Force pairs

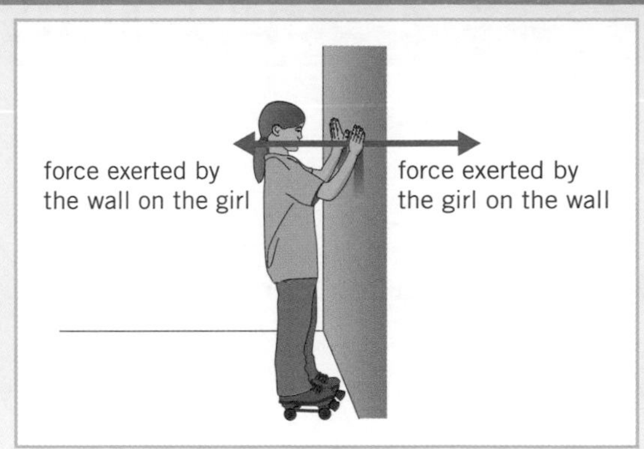

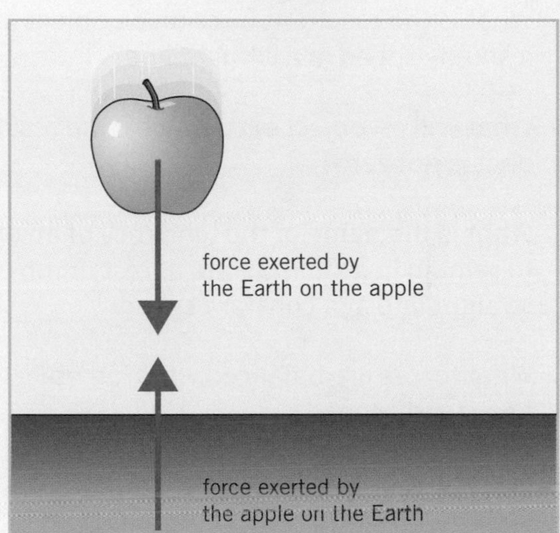

Revision tips

The formulas in physics can be used to answer text questions as well as calculation questions.

This often happens with questions about momentum or force.

Remember that any number multiplied by zero is zero, so the momentum of an object that is not moving is zero and the force from an object that is not accelerating is zero.

Momentum is conserved in explosions because:

- the total momentum before is zero
- the total momentum after is also zero because the different parts of the object travel in different directions and so the momentum of each part will cancel out with the momentum of another part.

If two moving objects **recoil** from each other, they start off with a total momentum of zero and end up moving away from each other with velocities v_1 and v_2. In this case, the law of conservation can be written as:

$$m_1v_1 + m_2v_2 = 0$$

Retrieval

Learn the answers to the questions below then cover the answers column with a piece of paper and write as many as you can. Check and repeat.

	P12 questions		Answers
1	What do we mean by inertia?	Put paper here	the tendency of an object to remain in a steady state (at rest or in uniform motion)
2	What does Newton's First Law say?		the velocity of an object will only change if a resultant force is acting on it
3	What is the resultant force on a stationary object?	Put paper here	zero
4	What is the resultant force on an object moving at a steady speed in a straight line?		zero
5	What will an object experience if the resultant force on it is not zero?		acceleration / change in velocity
6	What is the name of the tendency of an object to remain in a steady state at rest or moving in a straight line at a constant speed?	Put paper here	inertia
7	What forces are balanced when an object travels at a steady speed?		resistive forces = driving force
8	According to Newton's Second Law, what is the acceleration of an object proportional to?	Put paper here	the force acting on it
9	According to Newton's Second Law, what is the acceleration of an object inversely proportional to?		mass
10	What is the inertial mass of an object?	Put paper here	how difficult it is to change an object's velocity
11	What does Newton's Third Law say?		when two objects interact they exert equal and opposite forces on each other
12	Starting to move, stopping moving, speeding up, slowing down, and changing directions are all examples of which physical process?		acceleration / change in velocity

Now go back and use the questions below to check your knowledge from previous chapters.

Previous questions / Answers

	Previous questions	Answers
1	What is an alternating current (a.c.)?	current that repeatedly reverses direction
2	What is a scalar quantity?	only has a size (magnitude)
3	Which materials have low thermal conductivity?	thermal insulators
4	Which type of nuclear radiation is the most ionising?	alpha
5	Where does background radiation come from?	rocks, cosmic rays, fallout from nuclear weapons testing, nuclear accidents
6	What is the centre of mass?	the point through which the weight of an object can be considered to act
7	What is the unit for energy?	joule (J)

Put paper here

Put paper here

Required Practical

Force, mass, and acceleration

You need to be able to measure and explain how the acceleration of an object is linked to changes in force and mass.

For this practical, you can use two light gates to measure acceleration, or measure the time to travel a set distance. In the latter case the acceleration will be inversely proportional to the time.

To produce accurate and precise measurements you should:

- only change either the force or mass at one time
- use a video recording if not using light gates
- take repeat measurements and find the mean.

Worked example

A student sets up a trolley on a track. The trolley is attached to string, which goes over a pulley to a weight stack.

They set up light gates to record the acceleration as the force is changed.

Force in N	0	1	2	3	4	4
Acceleration in m/s²	0	4	8.1	12	16.2	20.2

1 Calculate the mass of the system.

force = mass × acceleration

$$\text{mass} = \frac{\text{force}}{\text{acceleration}}$$

for example, $\frac{1\,\text{N}}{4\,\text{m/s}^2} = 0.25\,\text{kg}$

2 Suggest how the student kept the mass of the system the same even though the number of weights on the stack increased.

They moved masses from the weight stack to the trolley.

3 What other force could be acting on the trolley, and how would this affect its acceleration?

Friction, which would cause the trolley's acceleration to be lower than expected.

Practice

A student connects a trolley to a piece of string attached to a weight stack. The weight stays the same, but the student increases the mass of the trolley.

She times how long it takes the trolley to travel 50 cm using a stop clock.

1 Sketch a graph of time against mass. Describe and explain the shape.

2 Describe how the student could ensure that the results are reproducible.

3 Suggest one benefit of using a larger distance.

Practice

Exam-style questions

01 A student wants to investigate the relationship between the mass of an object and its acceleration.

01.1 Describe an experiment that would enable the student to collect data to plot a graph of acceleration against mass. **[5 marks]**

01.2 Write down Newton's Second Law. **[1 mark]**

> **Exam Tip**
>
> It can be hard to remember which law was first, second, or third. Try looking at the first part of this question for clues.

01.3 Define the term inertia. **[1 mark]**

01.4 The student plotted their data on **Figure 1**. The student notices that a 0.2 kg mass does not have an acceleration of 20 m/s².

Figure 1

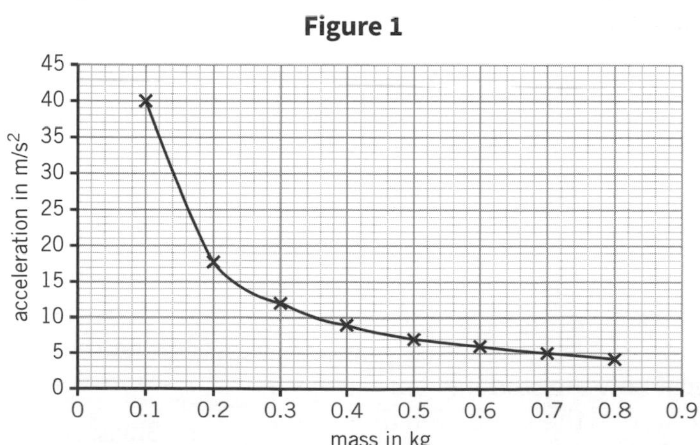

Suggest why the student expected that the acceleration would be 20 m/s².

Justify your answer. **[2 marks]**

01.5 Another student suggests that increasing the mass may have increased friction, which would have affected the acceleration.

Do you agree with this student? Explain your answer. **[3 marks]**

02 An ice hockey puck is moving along the ice at a steady speed.

02.1 Student **A** says: 'If the speed is steady, there are no forces acting on the puck'.

Do you agree with student **A**? Justify your answer. **[2 marks]**

Exam Tip

Thinking about Newton's Laws will help with these next few questions. Try writing them all down on the side here.

02.2 Student **B** says: 'The puck will only stop when the force it is carrying runs out'.

Do you agree with student **B**? Justify your answer. **[2 marks]**

02.3 One of the players hits the puck with a hockey stick and the puck moves in the opposite direction.

Student **C** says: 'While the stick is in contact with the puck, the stick exerts a bigger force on the puck than the puck exerts on the stick'.

Do you agree with student **C**? Justify your answer. **[2 marks]**

02.4 Determine whether the puck experienced an acceleration while in contact with the stick. Give reasons for your answer. **[2 marks]**

03 A company that transports food to supermarkets requires that lorry drivers are aware of how fast they are accelerating. The acceleration of a lorry should not exceed 1.5 m/s^2.

03.1 The engine of a lorry produces a force of 10 kN. Typical resistive forces are about 2 kN. A lorry and its load have a total mass of 8400 kg. A driver uses the maximum force possible. Determine whether the acceleration of her lorry exceeds the acceleration expected by the company. Justify your answer with a calculation. **[5 marks]**

Exam Tip

Be careful with the non-standard units here.

03.2 The lorry starts off fully loaded. The mass of the load is 3200 kg. The lorry stops at two supermarkets. It drops off half the load at each supermarket before going back to the depot. Calculate the acceleration of the lorry on its way to the second supermarket, and on its way back to the depot. Each time the driver uses the maximum force. Suggest whether this would be safe to do so. **[6 marks]**

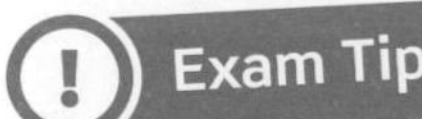

Exam Tip

You were given the mass of the lorry AND the load in the previous question. You have now been given the mass of just the load. The load will decrease at each supermarket. Read the question carefully to make sure you understand how it decreases.

04 A teacher demonstrates how changing the force applied to an object changes its acceleration. **Figure 2** shows a graph of the results. They draw a line of best fit.

Figure 2

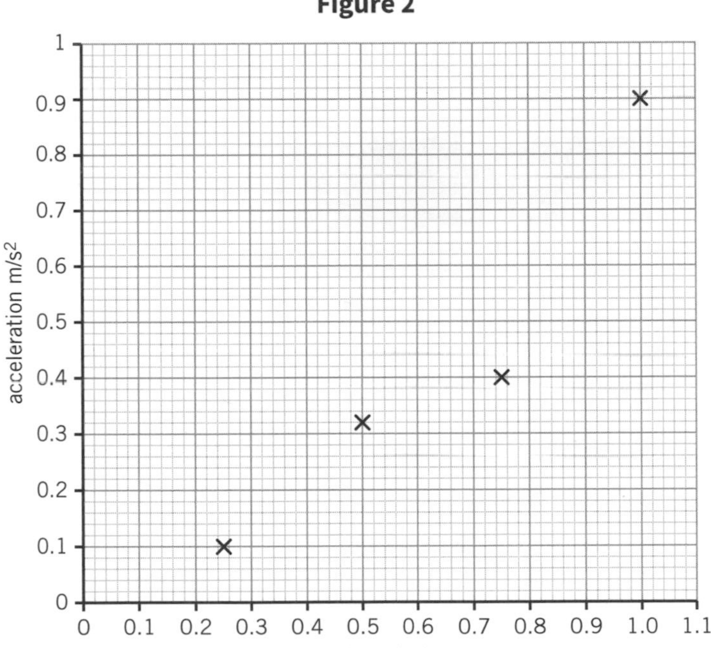

04.1 On **Figure 2** circle the measurement that is an outlier. **[1 mark]**

04.2 On **Figure 2** draw a line of best fit. **[1 mark]**

Exam Tip

Your line of best fit should best represent the data. This does not have to include all the points.

04.3 After the data has been plotted the teacher realises that the newtonmeter that she used does not read zero when there is no force applied. Identify the type of error that this problem produces. **[1 mark]**

04.4 Give the reason why you can tell that there is this type of error by looking at the graph. Use the graph to write down the reading on the newtonmeter when no force is applied. **[1 mark]**

04.5 Write down the equation that links force, mass, and acceleration **[1 mark]**

> **Exam Tip**
>
> Draw on your graph to ensure you use the correct values.

04.6 Use one of the points on your line of best fit to calculate the mass of the object. Explain how you have dealt with the error described in 04.3 and 04.4. **[4 marks]**

05 A car accelerates from 0 to 40 mph.

> **Exam Tip**
>
> Always show all of your working in a maths question. You can pick up some marks even if you don't get the final answer correct.

05.1 Convert 40 mph to m/s. There are 1609 m in 1 mile. There are 3600 s in an hour. **[3 marks]**

05.2 Write down what further information you need in order to state the velocity of the car. **[1 mark]**

05.3 Compare this speed to that of a typical cyclist. **[1 mark]**

> **Exam Tip**
>
> With any calculation, the first thing you should write down is the equation you are going to use.

05.4 The car accelerates over a distance of 30 m. Calculate the acceleration of the car. Assume the car is stationary at the start. Use an equation from the *Physics Equations Sheet*. Give your answer to an appropriate number of significant figures. **[5 marks]**

> **Exam Tip**
>
> When sketching this graph make sure you label 30 m and 60 m.

05.5 After accelerating for 30 m the car travels at a constant speed for 30 m. Sketch the velocity–time graph for this journey. **[2 marks]**

06 A theme park rollercoaster vehicle travels along a track. The forces acting on the vehicle are shown in **Table 1**.

Table 1

Section of straight track	Driving force in N	Resistive force in N
A	3000	2500
B	3000	3500
C	3500	3500

06.1 Write down the section of track where the vehicle is decelerating. **[1 mark]**

06.2 Write down the section of track where the vehicle is travelling at a steady speed. **[1 mark]**

06.3 At the end of the ride the only force acting on the vehicle is the resistive force due to the brakes. The mass of the vehicle and passengers is 3500 kg. Calculate the force required to stop the vehicle with a deceleration of 4 m/s^2. **[3 marks]**

06.4 On one section of the track a resultant force does not produce the change in speed predicted. Suggest why. **[1 mark]**

07 A student investigates friction. She ties string around a wooden block and applies a force. She measures the frictional force when the block just begins to move. She increases the weight on the block and repeats the experiment. She repeats the experiment with a brick.

The data is plotted in **Figure 3**.

Figure 3

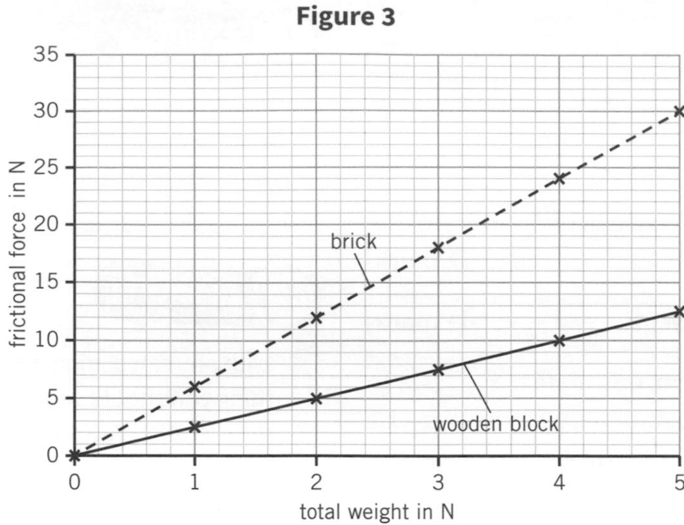

07.1 Write down the frictional force on a brick with a weight of 2.5 N when the brick is just about to move. **[1 mark]**

07.2 Write down the equation that links weight, mass, and gravitational field strength. **[1 mark]**

07.3 A wooden block has a weight of 1.4 N. Calculate the mass of the block. Gravitational field strength = 9.8 N/kg. **[3 marks]**

07.4 A student applies a force of 5 N to the wooden block. Calculate the expected acceleration of the block. **[6 marks]**

07.5 The actual acceleration is found to be larger than the value calculated in **07.4**. Suggest a reason why. **[1 mark]**

08 A rocket is launched from Earth towards the Moon. The rocket burns its fuel for 5 minutes and then turns off its engines. The rocket keeps moving through space at a speed of 10 km/s.

08.1 Write down the equation that links acceleration, velocity, and time. **[1 mark]**

08.2 Calculate the rocket's average acceleration during the time that the fuel burns. **[3 marks]**

08.3 Use the correct equation from the *Physics Equations Sheet* to calculate the distance that the rocket travels during this time. Give your answer in standard form. **[4 marks]**

08.4 The Moon is approximately 3.8×10^5 km from Earth. Assuming that the rocket continues at a speed of 10 km/s, calculate how long it will take to reach the Moon. Give your answer in an everyday unit. **[4 marks]**

Exam Tip

Standard form is a maths skill required for biology, chemistry, and physics. Make sure you are confident using standard form.

08.5 Suggest a reason why:

- it may take **less** time than you have calculated
- it may take **more** time than you have calculated. **[2 marks]**

09 A helicopter is collecting an object from a remote location. A rope, hanging down from the helicopter, is attached to the object. The object has a mass of 110 kg. The helicopter accelerates upwards with an acceleration of 2.0 m/s^2. Gravitational field strength = 9.8 N/kg.

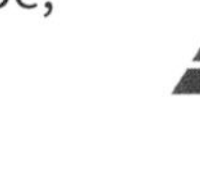

Exam Tip

There is a lot of information in the main body of this question. Write down a list of the key values:

- mass:
- acceleration:
- weight:
- force:

09.1 Calculate the weight of the object. **[3 marks]**

09.2 Calculate the force needed to accelerate the object at 2.0 m/s^2. **[3 marks]**

09.3 Show that the total force that the helicopter must exert on the object is about 1300 N. **[2 marks]**

09.4 Describe what happens to the object if the helicopter exerts a force of 1078 N. Give reasons for your answer. **[3 marks]**

10 A student wants to investigate the effect of force on the acceleration of a trolley. They place a trolley on a ramp, raise the ramp and the trolley accelerates. They use two light gates to measure the acceleration of the trolley.

Exam Tip

There are lots of different ways to do this practical. If you did it a different way in class then use what you did and apply it to this situation.

10.1 Describe how the student should use the light gates and trolley to measure the acceleration of the trolley. **[3 marks]**

10.2 The student decides to change the force acting on the trolley by raising the ramp to different heights. Explain how changing the height changes the force acting on the trolley. Include how the height affects the magnitude of the force. **[3 marks]**

10.3 The student performs their experiment. The results are shown in **Table 2**.

Table 2

Ramp height in cm	Acceleration in m/s^2
0	0.00
15	2.59
20	3.42
25	4.22
30	5.00
35	5.73
40	6.43

The student says 'because I used light gates, I only need to do the experiment once.' Do you agree? Justify your answer. **[2 marks]**

10.4 On **Figure 4** plot a graph of acceleration against ramp height. **[3 marks]**

Figure 4

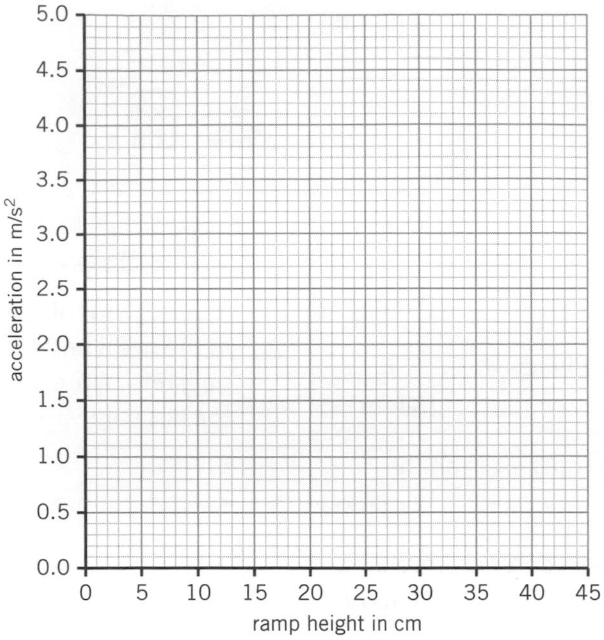

acceleration in m/s²: 0.0, 0.5, 1.0, 1.5, 2.0, 2.5, 3.0, 3.5, 4.0, 4.5, 5.0

ramp height in cm: 0, 5, 10, 15, 20, 25, 30, 35, 40, 45

Exam Tip

Always plot points using crosses and always draw a line of best fit.

10.5 The student continues to raise the ramp. Explain at what point there will be maximum possible acceleration in this experiment. Write down the value of maximum acceleration. **[2 marks]**

11 Some animals can accelerate very quickly. **Table 3** shows the acceleration and top speed of a leafhopper (an insect) and a cheetah.

Table 3

Animal	Acceleration in m/s^2	Top speed in m/s	Mass of animal
leafhopper	1000	4	2 mg
cheetah	5	30	50 kg

11.1 Write down the equation that links acceleration, force, and mass. **[1 mark]**

Exam Tip

In a show question, you know what the answer should be. So if you don't get that answer when you're doing your working out then you can try again until you get there. Just be careful of the amount of time you have in an exam.

11.2 Show that the force generated by the cheetah is 125 000 times greater than the force generated by the leafhopper. **[6 marks]**

11.3 Use **Table 3** to suggest whether the acceleration of the animal and the top speed of the animal are directly proportional. Justify your answer with calculations. **[4 marks]**

11.4 A company makes an acceleration suit for a human. The suit allows the wearer to have the same level of acceleration as a leafhopper. A human tries on the suit. They have a mass of 70 kg. Calculate the force produced by the suit. Compare your answer with the forces produced by vehicles in everyday road transport. Forces produced by cars are about 40 kN. **[4 marks]**

12 A person buying a car is comparing the performance of some cars. **Table 4** shows data from a magazine

Table 4

Car	Mass in kg	Time to go from 0 to 60 mph in s	Acceleration in m/s²
Tesla Model 3	1611	5.6	4.79
Audi A4	1565	7.1	3.78
BMW 440i	1864	4.8	5.59

12.1 They wonder whether the forces produced by the engines of the three cars are the same. Use **Table 4** to compare the forces. **[5 marks]**

This question is testing your data interpretation skills. Use the data from **Table 4** and refer to it in your answer.

12.2 They test the Audi. It travels at a steady speed of 50 mph with a driving force of 3 kN. Write down the resistive force acting on the car. **[1 mark]**

12.3 They take their foot off the accelerator pedal so that the only force acting on the car is the resistive force. Assume that the positive direction is in the direction of the driving force. Calculate the acceleration of the car. **[4 marks]**

13 The acceleration of a rocket is often used as an example of Newton's Third Law.

13.1 Write down Newton's Third Law. **[1 mark]**

13.2 When a rocket's engines fire, exhaust gases move downwards out of the rocket and the rocket accelerates upwards. Explain why there is an upwards force on the rocket. **[2 marks]**

13.3 Before the rocket launches it is stationary on the launch pad. A student says 'the forces acting on the rocket are equal and opposite. This is an example of Newton's Third Law.' Do you agree? Explain your answer. **[2 marks]**

14 A teacher sets up two tracks as shown in **Figure 5**. It is possible to roll a ball along each track without the ball coming off the track. The force of friction on both tracks can be ignored.

Figure 5

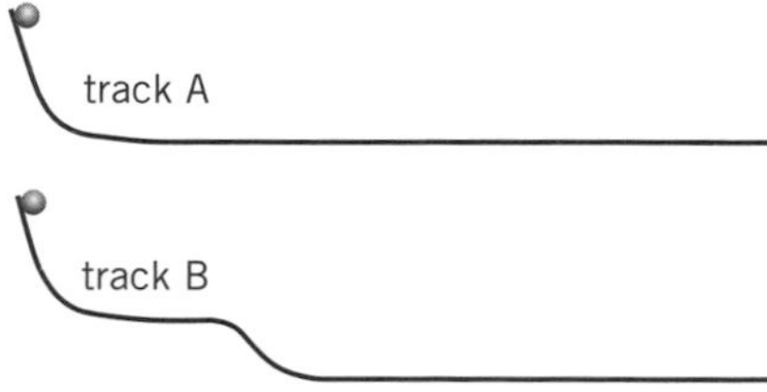

The teacher puts a ball on each track and releases them at the same time. Use ideas about forces and acceleration to suggest which ball reaches the end of its track first. Explain your answer. **[6 marks]**

Knowledge

P13 Braking and momentum

Newton's First Law

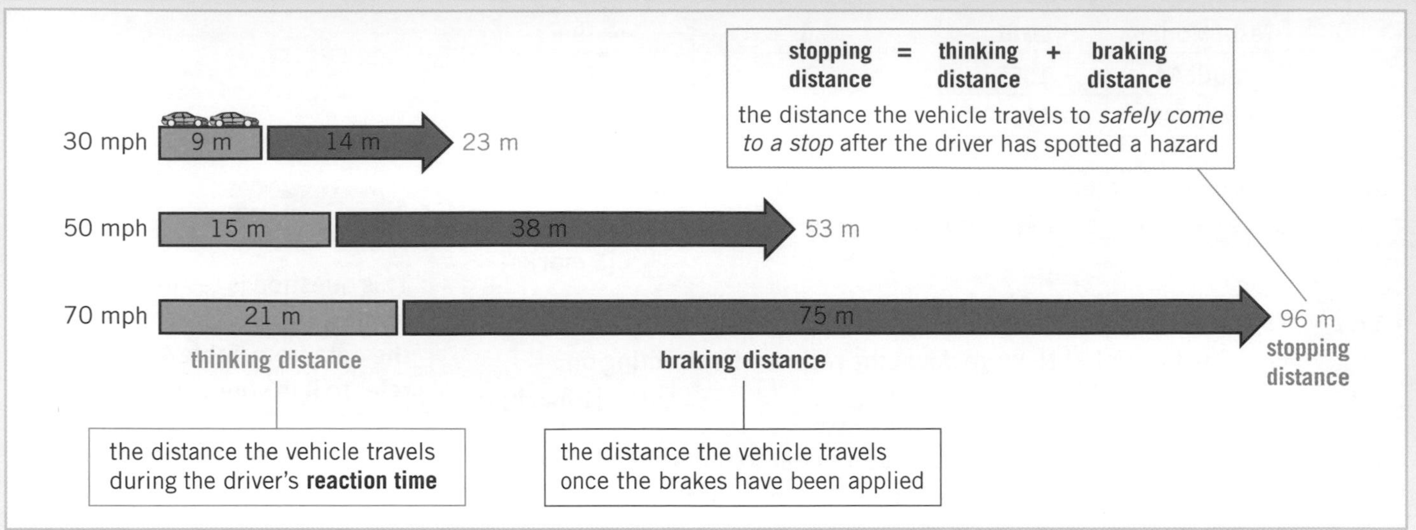

Speed has a bigger effect on braking distance than on thinking distance.

A graph of distance against speed for a vehicle shows that:

- thinking distance is proportional to the speed of the vehicle (straight line).
- braking distance is not proportional to speed but is proportional to the speed squared.

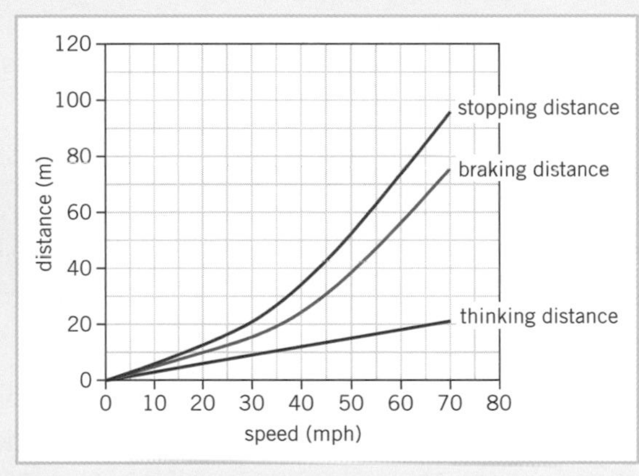

Factors affecting braking distance

The braking distance of a vehicle can be affected by:

- the speed of the vehicle
- road conditions
- the condition of brakes and tyres.

Any condition that causes less friction between the tyres and the road can lead to skidding, which increases the braking distance.

When the brakes of a vehicle are applied a frictional force is applied to its wheel.

Work done by the frictional force between the brakes and wheel transfers energy from the kinetic energy store of the car to the thermal energy stores of the brakes.

This increases the temperature of the brakes.

The braking force, braking distance, and energy transferred are related by the equation:

energy transfer (J) = *braking force* (N) × *distance* (m)

$$W = Fs$$

Key terms **Make sure you can write a definition for these key terms.**

braking distance deceleration reaction time stopping distance thinking distance

Reaction time

Reaction times vary from person to person, ranging from 0.2 s to 0.9 s.

Reaction time can be affected by:

- tiredness
- drugs
- alcohol
- distractions.

Reaction times can be measured in a number of ways, including:

Computer

A computer is used to time how long someone takes to respond to a sound or image on the screen.

Ruler drop test

The ruler is dropped between someone's fingers and the distance it falls before they catch it is used to calculate their reaction time.

Revision tips

Remember the unit for stopping distance is metres (m).

Don't get confused and read the unit m as miles.

The faster a vehicle moves or the greater its mass:

- the greater the amount of energy in its kinetic energy store
- the more work that has to be done to transfer the energy to slow it down
- the greater the braking force needed to stop it in a certain distance
- the greater the distance needed to stop it with a certain braking force.

Deceleration

The **deceleration** of the vehicle can be found using:

braking force (N) = mass (kg) × acceleration (m/s^2)

$$F = ma$$

The greater the braking force, the larger the deceleration.

If the braking force is very large, the brakes may overheat or the car may skid, causing the driver to lose control.

Changes in momentum

If an object is moving or is able to move, an unbalanced force acting on it will change its momentum.

Since $F = ma$ and $a = \frac{\Delta v}{t}$, we can write:

$$F = \frac{m\Delta v}{t}$$

where $m\Delta v$ is the change in momentum of an object.

The greater the time taken for the change in momentum of an object:

- the smaller the rate of change of momentum
- the smaller the force it experiences.

This means the force acting on an object is equal to the rate of change of momentum of the object.

Vehicle safety features increase the time taken for the change in momentum, e.g.:

- air bags, seat belts, and crumple zones in cars
- cycling helmets
- crash mats used for gymnastics

Revision tips

The m means mass, while the symbol for momentum is p.

Don't get these symbols mixed up!

Retrieval

Learn the answers to the questions below then cover the answers column with a piece of paper and write as many as you can. Check and repeat.

	P13 questions	Answers
1	What is the name given to the distance a vehicle travels to safely come to a stop after the driver has spotted a hazard?	stopping distance
2	What is thinking distance?	the distance vehicle travels during driver's reaction time
3	What is braking distance?	the distance vehicle travels once brakes have been applied
4	What is the relationship between stopping distance, thinking distance, and braking distance?	stopping distance = thinking distance + braking distance
5	Does the speed of a vehicle have a bigger effect on braking distance or thinking distance?	braking distance
6	Which distance is proportional to the speed of the vehicle?	thinking distance
7	Which distance increases by an increasing amount as speed increases?	braking distance
8	What are three factors that can affect the braking distance of a vehicle?	speed, road conditions, condition of tyres and brakes
9	What is the definition of one joule of work?	the work done when 1 N of force causes 1 m displacement
10	Why does the temperature of a vehicle's brakes increase when the brakes are applied?	work done by the frictional force between the brakes and the wheels transfers energy from the kinetic energy store of the car to the thermal energy store of the brakes
11	What can happen if the braking force used to stop a vehicle is very large?	brakes may overheat / the car may skid
12	What is the law of conservation of momentum?	in a closed system, the total momentum before an event is equal to the total momentum after it
13	What does $m\Delta v$ stand for?	change in momentum
14	How is the force acting on an object related to its momentum?	force acting on an object = rate of change of momentum
15	What are examples of everyday safety features which work by increasing the time taken for the change in momentum?	air bags, seat belts, crumple zones in cars, cycle helmets, crash mats in gyms, cushioned surfaces in children's playgrounds

Put paper here

Now go back and use the questions below to check your knowledge from previous chapters.

Previous questions

	Previous questions	Answers
1	What does Newton's Third Law say?	when two objects interact they exert equal and opposite forces on each other
2	What are the typical speeds for a person walking, running, and cycling?	1.5 m/s, 3.0 m/s, and 6.0 m/s respectively
3	What instrument can be used to measure the weight of an object?	calibrated spring-balance (newtonmeter)
4	What is irradiation?	exposing an object to nuclear radiation
5	What is the unit of pressure that is equal to one newton per square metre?	pascal (Pa)
6	On a graph of temperature against time for a substance being heated up or cooled down, what do the flat (horizontal) sections show?	the time when the substance is changing state and the temperature is not changing
7	Why do atoms have no overall charge?	they have equal numbers of positive protons and negative electrons

Put paper here

Put paper here

Put paper here

Maths Skills

Practise your maths skills using the worked example and practice questions below.

Significant figures

The **significant** numbers of a value are the ones which are meaningful based on what the number represents. We also use significant figures to make sure we are not introducing error by giving a false level of accuracy. Significant figures (s.f.) follow these rules:

- non zero digits are always significant
- zeros between non zero digits are significant
- leading zeros are not significant
- trailing zeros are not significant if there is no decimal point
- trailing zeros are significant if there is a decimal point in the number.

When answering questions you use the same number of s.f. as the data with the fewest s.f. in the question.

When giving a value to a specific number of s.f. you need to check if the value should be rounded. You use the next number after the desired s.f. to determine whether to round the value up or down.

Worked example

How many significant figures are there in 2.5070?

Answer: 5 s.f. as there are 3 non zero numbers, 1 zero between these, and 1 trailing zero after the decimal point.

Round 0.0037094 to 3 s.f.

Answer: Counting from the first s.f. to the number required gives you 0.00370 as leading zeros are not significant, but the zero between non zero digits is significant.

The next number (9) tells you to round the last s.f. up.

0.0037094 = 0.00371 to 3 s.f.

Practice

1 A How many significant figures are there in 1.2708?

2 How many significant figures are there in 0.0035008?

3 Round 35.073 to 3 s.f.

4 Round 67812 to 2 s.f.

Practice

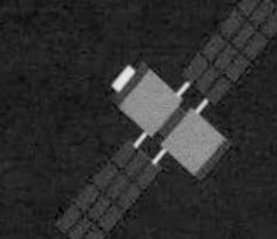

Exam-style questions

01 **Table 1** lists some factors that affect the thinking distance and braking distance of a car.

Table 1

Factor	Affects thinking distance	Affects braking distance
road conditions		
distractions in the car		
speed		

01.1 Tick **all** the correct boxes in the table. **[2 marks]**

01.2 Suggest **one** other factor that affects thinking distance.

Explain the effect of this factor on the thinking distance. **[3 marks]**

Exam Tip

Don't use factors that are mentioned in **01.1**. You are asked for another factor, so you won't gain a mark for repeating one.

01.3 Describe **one** type of road condition that can affect the stopping distance of a car.

Explain how and why the stopping distance is affected. **[3 marks]**

02 A student is learning about stopping distances in preparation for a driving test.

They produce **Table 2** in which to record stopping distances for different speeds.

Table 2

Speed in mph	Braking distance in m	Thinking distance in m	Stopping distance in m
30	13.9		
50			53.0

02.1 The Highway Code assumes that a reaction time is 0.67 s.

Calculate the thinking distance for a speed of 30 mph (13.4 m/s).

Write your answer to **two** significant figures. **[2 marks]**

Exam Tip

If you do not give your answer to two significant figures, you won't get full marks on this question.

______________ m

02.2 Calculate the stopping distance for 30 mph.
Write your answer to **two** significant figures. **[2 marks]**

______________ m

You need to use your answer from **02.1** to answer **02.2**.

If you got the wrong answer for **02.1** then the examiner can give you marks for 'error carried forward'. This is why you should always show your working. If you didn't get an answer for **02.1** at all then make one up and use that. The examiner can see that you've tried and you might still pick up some marks.

02.3 The thinking distance for a speed of 50 mph can be calculated using the formula:

thinking distance at 50 mph = thinking distance at 30 mph × $\frac{50}{30}$

Calculate the thinking distance at 50 mph.

Use the idea of proportion to justify the use of this calculation. **[3 marks]**

____

____

____

____

02.4 Calculate the braking distance at 50 mph. **[2 marks]**

______________ m

03 A student models the effect of road conditions on stopping distance using a trolley, a ramp, and different floor coverings (carpet, tiles, etc.).

03.1 Identify the independent variable, the dependent variable, and two control variables in this investigation. **[4 marks]**

There are four marks for **03.1**, so one mark for each variable. You can still pick up marks if you you can't work out all the variables.

03.2 Write an experimental method to obtain sufficient data to plot a graph. **[5 marks]**

03.3 Identify a source of error in the experiment. Suggest an improvement to reduce its effect. **[2 marks]**

03.4 Evaluate to what extent this is a good model of the effect of surface on the stopping distance of a car. Use ideas about work and friction in your answer. **[3 marks]**

03.5 Suggest a limitation of this model. Give a reason for your answer. **[2 marks]**

04 A teacher is explaining thinking distance to a physics class. In pairs, the students measure their reaction time by grabbing a falling ruler. In each pair, student **A** holds a ruler just above student **B**'s hand. Student **A** then drops the ruler and student **B** grabs it. They record the distance the ruler has fallen. **Figure 1** shows the relationship between the distance fallen and reaction time.

Figure 1

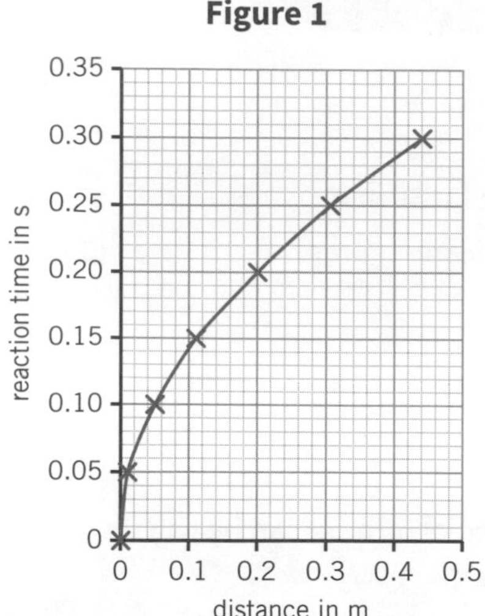

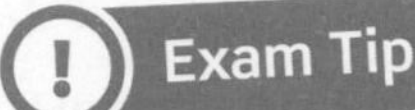

You may be familiar with this from biology required practicals, but you need to think about it in a physics context. Don't let the change in context put you off.

04.1 Explain the relationship between reaction time and distance shown on **Figure 1**. **[2 marks]**

04.2 One student measures a distance on the ruler of 25 cm. Use **Figure 1** to work out the reaction time of the student. **[1 mark]**

04.3 Explain whether the reaction time of the student falls within the range of normal reaction times. **[2 marks]**

04.4 Suggest **one** source of error in doing this experiment. Explain the possible effect of this error on the measurement of reaction time. **[3 marks]**

05 A medical journal published an article on injuries due to falling coconuts. Here is a summary of the study as reported in the journal.

"A 4-year review of trauma admissions to the Provincial Hospital, Alotau, Papua New Guinea, reveals that 2.5% of such admissions were due to being struck by falling coconuts. Since mature coconut palms may have a height of up to 35 meters and an unhusked coconut may have mass of up to 4 kg, blows to the head of a force exceeding 1 metric tonne, or 1000 kg, are possible."

05.1 When the scientists completed their investigation, they wrote a paper and submitted it to the journal. Describe what happens to an article before it is published. Give the name of this process. **[2 marks]**

05.2 Suggest why the statement *a force exceeding 1 metric tonne* is incorrect. **[1 mark]**

05.3 Write down the equation that links gravitational field strength, mass, and weight. **[1 mark]**

05.4 Calculate the force exerted by a mass of 1000 kg. Gravitational field strength = 9.8 N/kg. **[2 marks]**

Exam Tip

Sort out which information is useful and what is just a distraction by crossing out information that will not help you answer the question. For example, the fact that this happened in a Provincial Hospital, Alotau, Papua New Guinea is not helpful for the physics. This can be crossed out. However, the height that the coconuts fall is useful for the physics.

05.5 Suggest why the authors chose to describe the force in this way in the summary. **[1 mark]**

05.6 A coconut of mass 2.5 kg falls from a tree in 2.5 seconds. Calculate the momentum of the coconut just before it hits the ground. **[4 marks]**

06 Two people are on an ice rink. They are wearing ice skates. Person **A** is stationary. Person **B** is skating towards Person **A** with a speed of 5.0 m/s to the right. They collide. The mass of person **B** is 65 kg.

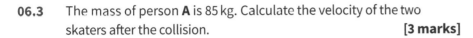

Exam Tip

It can help to draw a diagram to show the problem. Draw arrows to show direction and label each direction as positive or negative.

06.1 Calculate the momentum of person **B** before the collision. Give the unit with your answer. **[3 marks]**

06.2 When they collide, the people link arms and move off together without stopping. Explain why the velocity of the skaters is different after the collision. Write down an assumption that you have made in your answer. **[4 marks]**

06.3 The mass of person **A** is 85 kg. Calculate the velocity of the two skaters after the collision. **[3 marks]**

06.4 At the end of the session the two skaters skate to the edge of the rink, collide gently with the barrier, and stop. Explain why momentum is not conserved in this situation. **[2 marks]**

07 A driver is on the motorway travelling at 70 mph.

07.1 Convert 70 mph to m/s. There are 1609 m in 1 mile. **[3 marks]**

Exam Tip

The chevrons are painted at set distances apart.

07.2 The driver sees a sign that says:
'*keep two chevrons apart*'.
Chevrons are large '/\' shapes painted on the road. Each driver should be able to see two chevrons between themselves and the driver in front of them. Suggest what this system is designed to do. **[2 marks]**

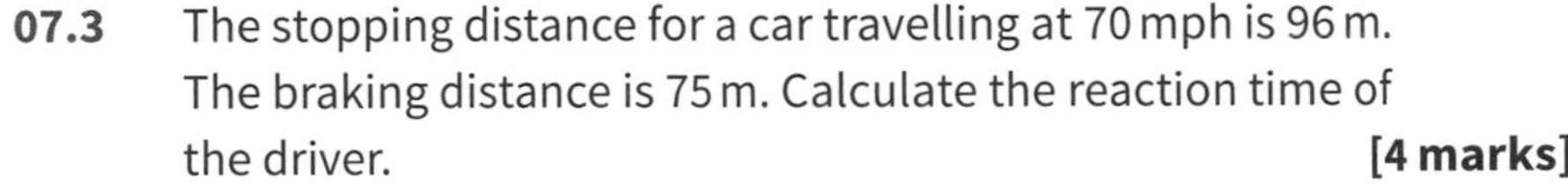

07.3 The stopping distance for a car travelling at 70 mph is 96 m. The braking distance is 75 m. Calculate the reaction time of the driver. **[4 marks]**

07.4 A student calculates that the distance between the chevrons should be 48 m. Suggest how the student worked out this number. **[1 mark]**

07.5 The chevrons are actually 40 m apart. Suggest a reason why the actual distance is smaller than the value in **07.4**. **[2 marks]**

08 It is very important to be able to stop a road vehicle quickly in an emergency. **Figure 2** shows a graph of braking distance against speed for a car in normal conditions and in an emergency.

Figure 2

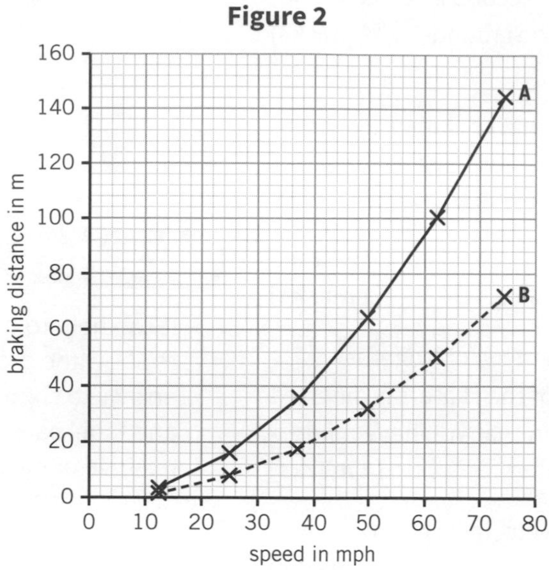

08.1 Write down which curve on **Figure 2**, **A** or **B**, relates to an emergency stop. Use the words force and acceleration to explain your answer. **[4 marks]**

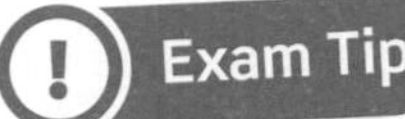

The question tells you some key words to use so make sure that you do use them.

08.2 Suggest **one** harmful effect of having to do an emergency stop. **[1 mark]**

08.3 **Figure 2** does not include the thinking distance. Compare the thinking distance when stopping in normal conditions and when doing an emergency stop. **[2 marks]**

09 **Figure 3** shows the effect that using a mobile phone has on reaction times.

Figure 3

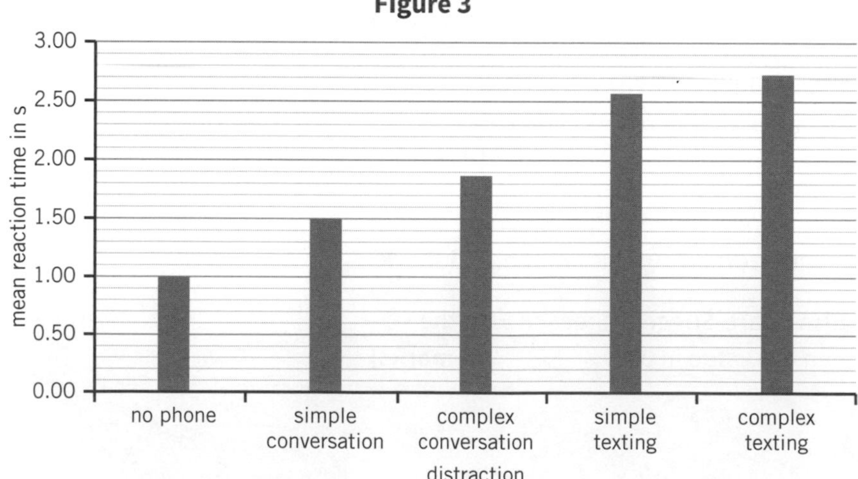

09.1 Many countries have now banned the use of mobile phones by the drivers of cars. Use **Figure 3** to suggest why. Explain your reasoning. **[3 marks]**

Exam Tip

Use data from the graph and compare the different uses of mobile phones to the no phone column.

09.2 A car is travelling at 30 mph (13.4 m/s). Calculate the difference in thinking distance between a driver holding a complex conversation and a driver with no phone. **[3 marks]**

09.3 Calculate the speed that the person with no phone would need to travel at to have the same thinking distance as the person having the complex conversation at 30 mph. Use ratios to convert the speed to miles per hour.

Comment on your answer. **[5 marks]**

Exam Tip

You'll need to decide if you are going to use the 30 mph or the 13 m/s as the speed. Look at the graph to help you decide.

10 A company is writing an advice leaflet to go with the baby bouncer that they sell. The leaflet will give parents a guide to the height off the floor to fix the bouncer for babies of different weights.

Figure 4

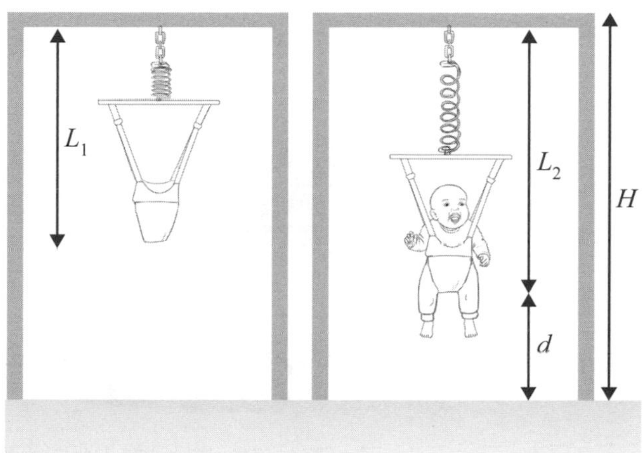

For safety, the distance d as shown in **Figure 4** should be no less than 50 cm. The length of the chain, spring, and harness L_1 = 1.00 m. When the baby is put in the harness, the length L_2 = 1.25 m.

The weight of the baby is 100 N.

10.1 Write down the equation that links force, spring constant, and extension. **[1 mark]**

10.2 Calculate the spring constant of the spring holding the harness. **[4 marks]**

10.3 Calculate the height H, from the floor that you need to fix the baby bouncer so the baby can be safe. **[2 marks]**

Exam Tip

You can write down the word or the symbol equation – however you best remember things.

11 A company has been experimenting with airbags. They have used a forcemeter connected to a datalogger to monitor the force on a model person in a car. The car collides with a barrier. In one test there is no air bag attached to the model. In the second test, the air bag inflates.

The graph of force on the model person against time for both collisions is shown in **Figure 5**.

> **Exam Tip**
>
> Label these lines 'first test' and 'second test'. This might make things clearer for you.

Figure 5

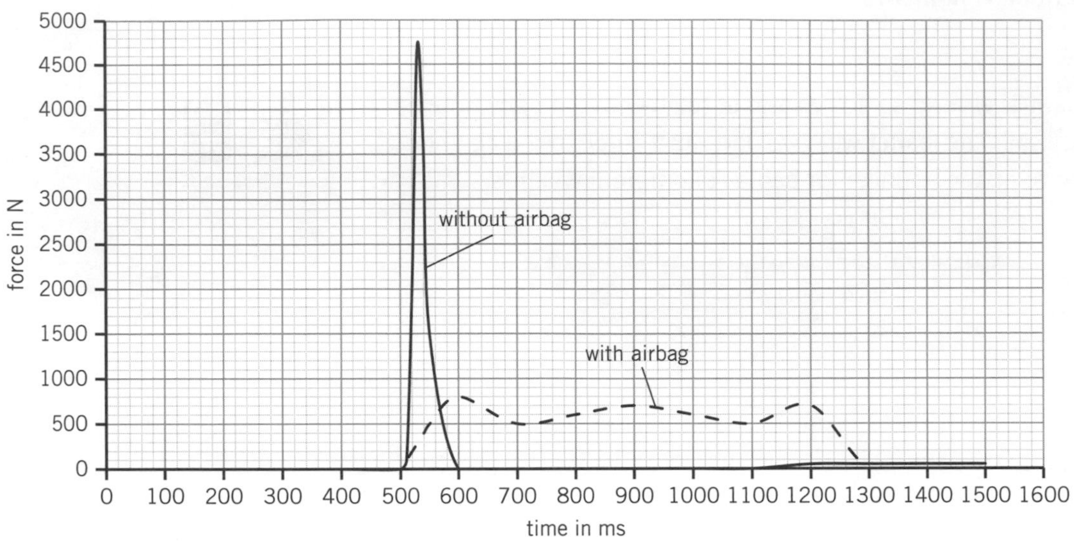

11.1 Use **Figure 5** to compare the maximum force experienced by the model person with and without an airbag. **[3 marks]**

11.2 Use **Figure 5** to estimate the average force exerted by the airbag while it was bringing the model person to a stop. **[1 mark]**

11.3 Calculate the initial velocity of the model person when an airbag is used. Assume that the time the airbag took to inflate was 750 ms. Assume that the mass of the model person is 60 kg. Use the equation momentum = mass × velocity and an equation from the *Physics Equation Sheet*. **[5 marks]**

11.4 Explain why airbags result in less injury to the human body involved in a car accident. **[2 marks]**

> **Exam Tip**
>
> Draw lines across for the top of the impact line to the y-axis to help you read of the maximum force values

> **Exam Tip**
>
> Here, you have to merge the equation you have been given with one from the *Physics Equation Sheet*.

12 The driver of a car reacts to a car braking suddenly in front of them. The car is initially travelling at 45 mph (20 m/s) and comes to a complete stop.

12.1 Describe the process by which energy is transferred from a kinetic energy store to a thermal energy store. **[1 mark]**

12.2 When energy is transferred to a thermal energy store, objects get hot. Write down which objects get hotter when the driver brakes. **[1 mark]**

12.3 The speed of the car changes over a time of 4.3 seconds. Calculate the deceleration of the car. Give **one** assumption that you make when you do this calculation. Suggest whether this assumption is likely, or not likely, to be correct. Explain your answer. **[5 marks]**

> **Exam Tip**
>
> You make assumptions for most calculations even if you don't realise it. Think about the motion of a car and what assumption you could have made.

12.4 The mass of the car is 1250 kg. Calculate the braking force. Compare your answer with the typical driving force of a car. **[3 marks]**

13 A teacher demonstrates what is meant by 'work' in science. They lift a 1 kg mass a distance of 1 m.

13.1 Use the equation that links weight, mass, and gravitational field strength to calculate the force needed to lift the mass. Gravitational field strength is 9.8 N/kg. **[3 marks]**

13.2 Use the equation that links work, force, and distance to calculate the work done. **[3 marks]**

13.3 The teacher holds the mass without moving it. Write down the force the teacher uses to hold the mass. **[1 mark]**

14 A local council wants to reduce the number of injuries due to cars striking pedestrians. The council wants to change the speed limit from 30 mph (13.4 m/s) to 25 mph (11.2 m/s) in all built-up areas. To persuade the public to accept the new speed limit, they are considering two approaches.

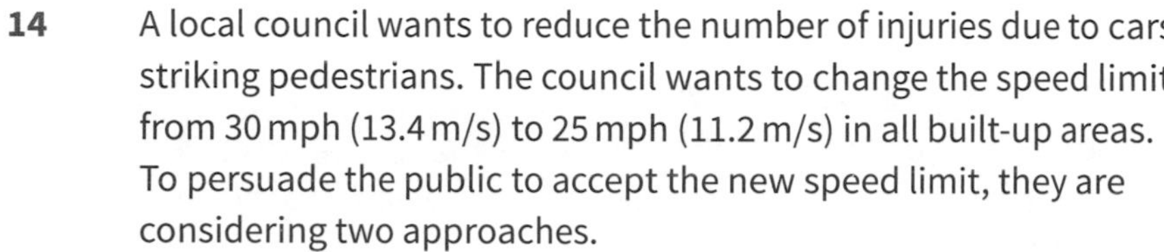

Approach 1: Explain how much more distance you need to stop at a higher speed.

Approach 2: Explain how fast a car will be moving if it collides with a pedestrian because it is moving too fast and is unable to stop.

A car travelling at 25 mph comes to a complete stop after a distance of 9.4 m.

14.1 The mass of a car is 1500 kg. Calculate the kinetic energy of the car as it comes to a complete stop when travelling at 30 mph. Give your answer in standard form. **[3 marks]**

14.2 Using your answer from **05.1**, calculate the difference in the distance the car travels to come to a complete stop when travelling at 30 mph compared to when travelling at 25 mph. The braking force of a car is 10 kN. **[5 marks]**

14.3 A car travelling at 30 mph with an acceleration of $-6.7\ m/s^2$ will still be moving after travelling 9.4 m. Use ideas about deceleration and final speed to calculate how fast a car travelling at 30 mph would still be moving after travelling 9.4 m. Use an equation from the *Physics Equation Sheet*. Convert your answer to miles per hour. 1 mile = 1609 m **[5 marks]**

14.4 Suggest which approach is likely to be more successful in persuading the public to adopt a lower speed limit. **[2 marks]**

Exam Tip

Write down which equation you will use to answer **14.1**. Selecting the correct equation is an important skill that you need to show in the exams.

Exam Tip

There are lots of different units in **14.3**. Make sure you don't mix up m from meters with the m in miles per hour. It would be easy to get confused and not get the marks.

Knowledge

P14 Mechanical waves

Waves in air, fluids, and solids

Waves transfer energy from one place to another without transferring matter. Waves may be **transverse** or **longitudinal**.

For waves in water and air, it is the wave and not the substance that moves.

- When a light object is dropped into still water, it produces ripples (waves) on the water which spread out, but neither the object nor the water moves with the ripples.
- When you speak, your voice box vibrates, making sound waves travel through the air. The air itself does not travel away from your throat, otherwise a vacuum would be created.

Mechanical waves require a substance (a medium) to travel through.

Examples of mechanical waves include sound waves, water waves, waves on springs and ropes, and seismic waves produced by earthquakes.

When waves travel through a substance, the particles in the substance **oscillate** (vibrate) and pass energy on to neighbouring particles.

Transverse waves

The oscillations of a transverse wave are *perpendicular* (at right angles) to the direction in which the waves transfer energy.

Ripples on the surface of water are an example of transverse waves.

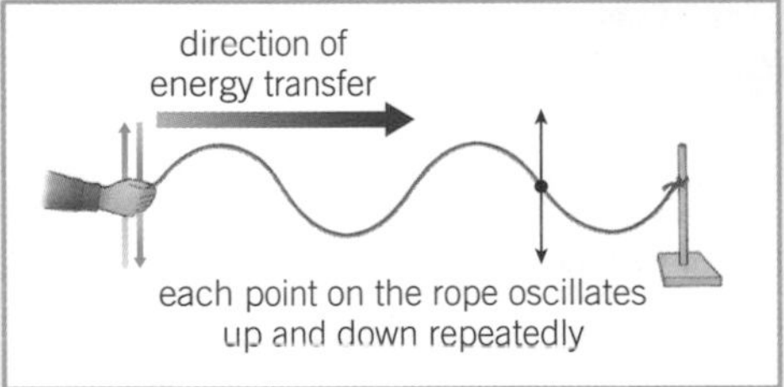

Longitudinal waves

The oscillations of a longitudinal wave are *parallel* to the direction in which the waves transfer energy.

Longitudinal waves cause particles in a substance to be squashed closer together and pulled further apart, producing areas of **compression** and **rarefaction** in the substance.

Sound waves in air are an example of longitudinal waves.

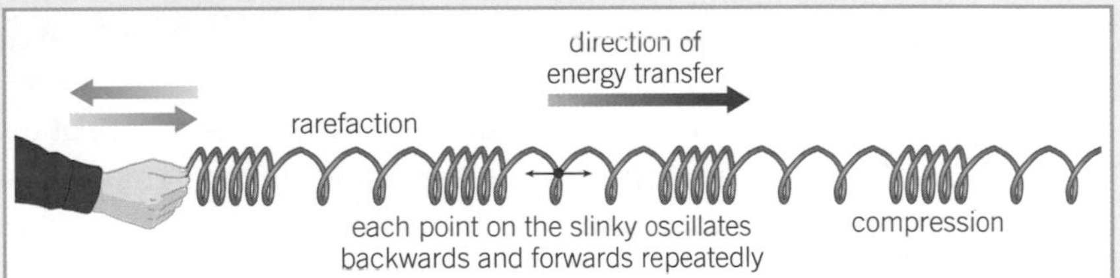

Wave motion is described by a number of properties.

Property	Description	Unit
amplitude A	maximum displacement of a point on a wave from its undisturbed position	metre (m)
frequency f	number of waves passing a fixed point per second	hertz (Hz)
period T	time taken for one complete wave to pass a fixed point	second (s)
wavelength λ	distance from one point on a wave to the equivalent point on the next wave	metre (m)
wave speed v	distance travelled by each wave per second, and the speed at which energy is transferred by the wave	metres per second (m/s)

Key terms **Make sure you can write a definition for these key terms.**

absorption amplitude compression frequency incidence longitudinal mechanical wave oscillate

Properties of waves

Frequency and period are related by the equation:

$$\text{period (s)} = \frac{1}{\text{frequency (Hz)}} \qquad T = \frac{1}{f}$$

All waves obey the wave equation:

wave speed (m/s) = frequency (Hz) × wavelength (m)

$$v = f\lambda$$

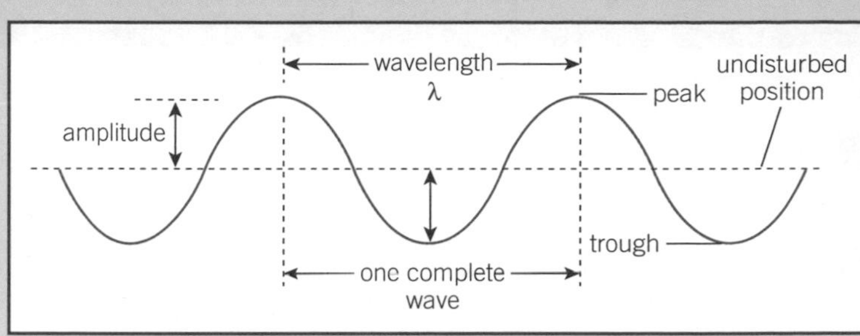

When waves travel from one medium to another, their speed and wavelength may change but the frequency always stays the same.

The speed of ripples on water can be slow enough to measure using a stopwatch and ruler, and applying the equation:

$$\text{speed (m/s)} = \frac{\text{distance (m)}}{\text{time (s)}}$$

The speed of sound in air can be measured by using a stopwatch to measure the time taken for a sound to travel a known distance, and applying the same equation.

Reflection of waves

When waves arrive at the boundary between two different substances, one or more of the following things can happen:

Absorption – the energy of the waves is transferred to the energy stores of the substance they travel into (for example, when food is heated in a microwave)

Reflection – the waves bounce back

Refraction – the waves change speed and direction as they cross the boundary

Transmission – the waves carry on moving once they've crossed the boundary, but may be refracted

Ray diagrams can be used to show what happens when a wave is reflected at a surface.

To correctly draw a ray diagram for reflection:

1 use a ruler to draw all lines for the rays
2 draw a single arrow on the rays to show the direction the wave is travelling
3 draw a dotted line at right angles to the surface at the point of **incidence** (this line is normal to the surface)
4 label the normal, angle of incidence (i), and angle of reflection (r).

When reflection happens at a surface, the angle of incidence is always equal to the angle of reflection:

$$i = r$$

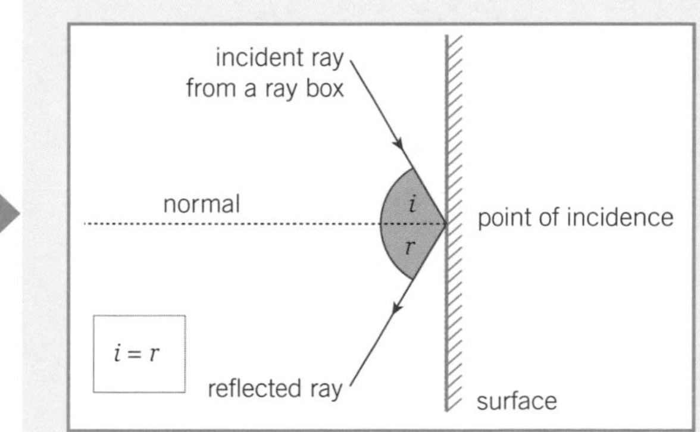

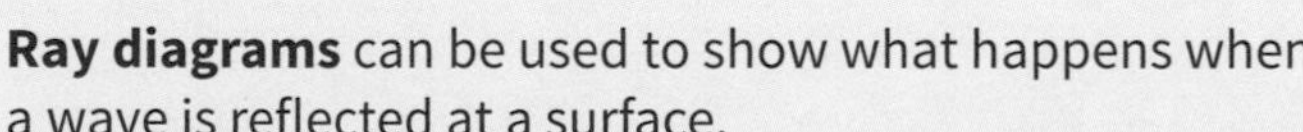

period ray diagram reflection rarefaction transmission transverse wavelength wave speed

Retrieval

Learn the answers to the questions below then cover the answers column with a piece of paper and write as many as you can. Check and repeat.

	P14 questions	Answers
1	What is a transverse wave?	oscillations/vibrations are perpendicular (at right angles) to the direction of energy transfer
2	What is a longitudinal wave?	oscillations/vibrations are parallel to the direction of energy transfer
3	Give an example of a transverse wave.	electromagnetic waves
4	Give an example of a longitudinal wave.	sound waves
5	What is a compression?	area in longitudinal waves where the particles are squashed closer together
6	What is a rarefaction?	area in longitudinal waves where the particles are pulled further apart
7	What is the amplitude of a wave?	maximum displacement of a point on the wave from its undisturbed position
8	What is the wavelength of a wave?	distance from a point on one wave to the equivalent point on the adjacent wave
9	What is the frequency of a wave?	number of waves passing a fixed point per second
10	What unit is frequency measured in?	hertz (Hz)
11	What property of a wave always stays the same when it travels from one medium to another?	frequency
12	What rule do waves follow when they reflect off a surface?	angle of incidence = angle of reflection
13	What happens when waves are transmitted at a boundary between two substances?	they carry on moving at a different speed
14	What happens when waves are absorbed by a substance?	energy of the wave is transferred to energy stores of the substance

Put paper here

Now go back and use the questions below to check your knowledge from previous chapters.

Previous questions / Answers

	Previous questions	Answers
1	What do we mean by inertia?	the tendency of an object to remain in a steady state (at rest or in uniform motion)
2	What is acceleration?	change in velocity of an object per second
3	What is the unit of acceleration?	m/s^2
4	What is nuclear fission?	the splitting of a large and unstable nucleus into two smaller nuclei
5	How does the height of the atmosphere compare to the radius of the Earth?	it is smaller

Put paper here

Put paper here

Required Practical

Practise answering questions on the required practicals using the example below. You need to be able to apply your skills and knowledge to other practicals too.

Investigating waves

You need to be able to measure the frequency and wavelength of waves through a liquid and a solid. A ripple tank will be used to make water waves, and a vibration generator to make standing waves on a string.

To be accurate and precise in your investigation you need to:

- count and measure the length of multiple water waves
- measure the length of multiple waves on a string
- repeat your measurements and find the mean.

You should be able to evaluate the suitability of the equipment used to measure wave speeds in liquids and solids.

Worked example

A student sets up a ripple tank.

A lamp above the tank produces this image of the waves on a piece of paper below the tank.

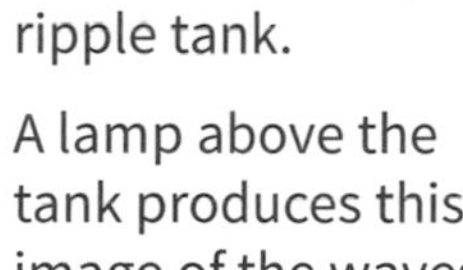

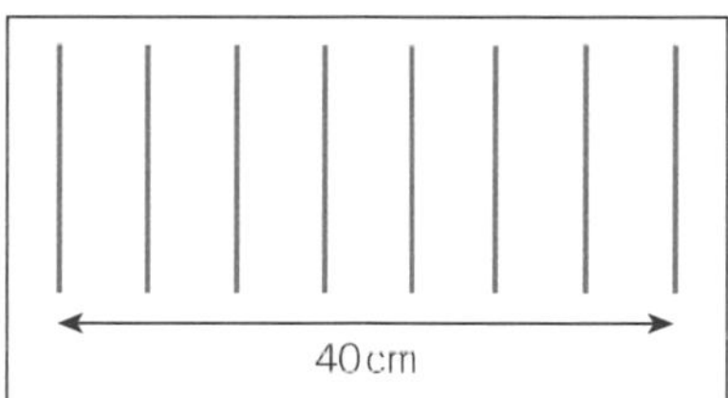

1 Calculate the wavelength of the waves.

Number of waves = 7

wavelength $= \frac{40}{7} = 5.7\,\text{cm} = 0.057\,\text{m}$

2 The student counts 25 waves reaching the end of the tank in 20 seconds. Calculate the frequency of the waves.

$\text{frequency} = \frac{\text{number of waves}}{\text{time}}$

$\frac{25}{20} = 1.25\,\text{Hz} = 1.3\,\text{Hz}$ to 2 s.f.

3 Calculate the speed of the waves.

speed = frequency × wavelength

$1.3 \times 0.057 = 0.0741\,\text{m/s} = 0.074\,\text{m/s}$ to 2 s.f.

4 Suggest one improvement the student could make to improve the accuracy of their speed measurement.

Repeat all measurements of wavelength and frequency and take the average.

Practice

A teacher connects a vibration generator to a piece of elastic.

She changes the frequency of the vibration generator and measures the wavelength of the elastic, producing the data below.

Frequency in Hz	Wavelength in m
4	4.86
8	2.51
12	1.25
16	1.29
20	1.02

1 Plot a graph of these data. Identify the outlier.

2 Use the data to calculate the average wave speed in m/s. Show your working.

Practice

Exam-style questions

01 A student makes a wave on a long thin spring.

The spring is attached to the wall.

The wave is shown in **Figure 1**.

Figure 1

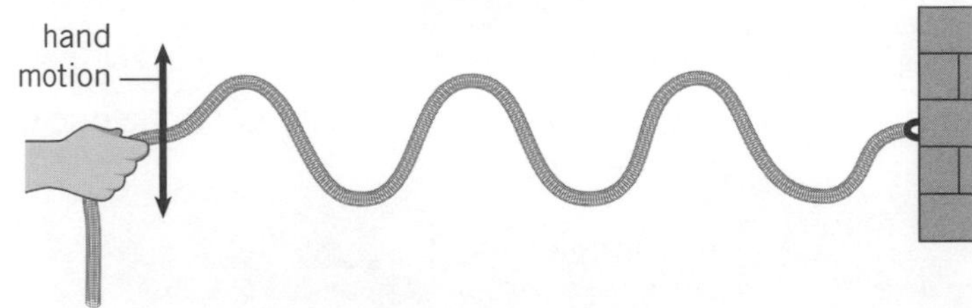

01.1 Write down the type of wave that the student produces. **[1 mark]**

____

01.2 On **Figure 1** draw an arrow to show the wavelength of the wave. **[1 mark]**

01.3 Describe what the student should do to produce waves with a smaller amplitude.

Explain your answer. **[2 marks]**

____

____

____

02 One of the highest measured ocean waves was measured at a height of 34 metres from peak to trough.

The period of the wave was 14.8 s.

The wavelength was calculated to be 342 m.

02.1 Calculate the amplitude of the wave. **[2 marks]**

Amplitude = ____________ m

02.2 Calculate the speed of the wave.

Use an equation from the *Physics Equations Sheet*.

Show your working. **[4 marks]**

Speed = ____________ m/s

02.3 A student compares this wave with the waves seen in a ripple tank. A ripple tank is a tray of water on four legs.

Estimate the amplitude of waves in a ripple tank. **[1 mark]**

____

02.4 The speed of waves in a ripple tank is found to be 50 cm/s. Suggest whether the speed of a wave is proportional to its amplitude. **[3 marks]**

____

____

____

____

Exam Tip

Look out for non-standard units.

Exam Tip

The height was given in the question stem – 34 metres from peak to trough.

03 A child is throwing stones into a pond. The ripples move across the surface of the water.

03.1 Compare the motion of the surface of the water with the motion of the wave. **[1 mark]**

03.2 The stone makes a sound when it hits the water. The sound wave and the ripple move towards the child. Describe the difference between the motion of the particles on the surface of the water and the motion of the air particles in the sound wave. **[2 marks]**

03.3 Write down the equation that links wave speed, frequency, and wavelength. **[1 mark]**

03.4 The frequency of the sound that the stone makes is 400 Hz. The speed of sound in air is 340 m/s. Calculate the wavelength of the sound waves. **[3 marks]**

04 You can use a slinky coil to model waves. The wave on the slinky coil has compressions and rarefactions.

04.1 On **Figure 2** write the letter **C** above a compression, and the letter **R** above a rarefaction. **[2 marks]**

04.2 Show that the wavelength of the wave is 0.5 m. **[2 marks]**

Figure 2

04.3 Write down the equation that links wave speed, frequency, and wavelength. **[1 mark]**

04.4 The wave is travelling at 1.0 m/s. Calculate the frequency of the wave. State the unit. Explain what this means in terms of the motion of the person's hand. **[5 marks]**

05 A student watches a video about different types of wave. The video describes ripples on a water surface and sound waves.

05.1 Write down which of these types of wave is longitudinal. **[1 mark]**

05.2 Describe an experiment to show that, as the wave travels, the air and water particles are not carries with it. Describe the observations that you would make and what those observations show. **[6 marks]**

06 A scientist attempted to measure the speed of sound experimentally by measuring the time difference between spotting the flash of a gun and hearing the sound produced by the gun. The experiment was carried out over a long distance on a day without any wind. The value obtained was 478.4 m/s.

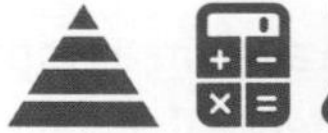

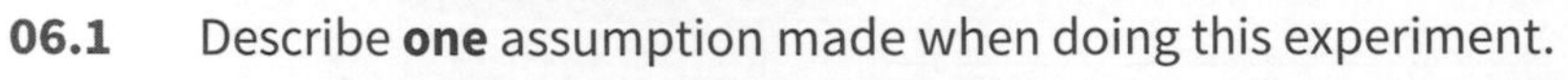

06.1 Describe **one** assumption made when doing this experiment. **[1 mark]**

06.2 Describe the measurements made and how they were used to calculate the speed of sound. **[2 marks]**

06.3 The accepted speed of sound is 340 m/s. Calculate the percentage difference between the accepted value and the value first calculated. Show your working. **[3 marks]**

06.4 Another scientist stood 29 km away from a canon. The scientist measured the time interval between hearing the cannon being fired and the scientist hearing the canon. They calculated the speed of sound as 332 m/s. Calculate the time interval between hearing the sound from the cannon. Use the speed of sound calculated by the scientist. **[4 marks]**

06.5 Suggest why the difference between the measured and accepted value of the speed of sound calculated in this experiment is much smaller than for the first scientist's experiment. **[2 marks]**

07 A student sets up a ripple tank to measure the speed of water waves. The student takes a bar with a motor attached to it to make it vibrate. The bar is attached to a power supply and is partly submerged in a tray of water. A lamp above the tray of water projects an image of the ripples onto the desk below the ripple tank.

07.1 Describe a hazard associated with this experiment. Describe a strategy to reduce the risk of injury due to the hazard. **[2 marks]**

Exam Tip

This is one of the required practicals. If you can't remember doing it try looking it up in your notes.

07.2 The student turns on the motor and sees the image of ripples moving across the desk. The student records that the number of waves passing a point in 10 s is 5 and that the number of waves in 20 cm on the desk is 15. Suggest what can be calculated from this data. Include the units. **[2 marks]**

07.3 Describe how to perform the calculations you have suggested using the data collected by the student. Explain your reasoning. **[4 marks]**

07.4 Write down the equation that links wave speed, frequency, and wavelength **[1 mark]**

07.5 Use the data recorded by the student to calculate the speed of the ripples in the tank. Give your answer in standard form. Explain why the answer should be given to one significant figure. **[6 marks]**

You should always match the resolution of your answer to the data given in the question.

08 A wave in a ripple tank is reflected from a barrier.

08.1 Draw a diagram to show the reflection of a wave at a barrier at an angle of 30° to the normal. **[4 marks]**

08.1 has given you a specific angle to draw. That means you need to get your protractors out and draw it.

08.2 Explain why the amplitude of the wave that hits the barrier is bigger than the amplitude of the wave reflected from the barrier. **[2 marks]**

08.3 Sound is a wave. Sound waves can be reflected from barriers, such as walls and windows. Suggest **one** observation that you have made that shows that sound can also be transmitted through barriers. **[1 mark]**

08.4 A sound wave is transmitted through a concrete wall. The velocity of the sound wave increases. Write down what happens to the frequency and wavelength of the sound wave as it moves from the air into the wall. **[2 marks]**

09 A driver is driving on an icy road.

09.1 The stopping distance of a car is affected by the condition of the road surface. Explain why. **[2 marks]**

09.2 The stopping distance is also affected by speed. Explain why speed affects thinking distance and braking distance. **[4 marks]**

09.3 **Figure 3** shows the stopping distances at different speeds on an icy surface and a dry road. Suggest which curve, **A** or **B**, is the graph for an icy surface. Explain your answer. **[2 marks]**

Figure 3

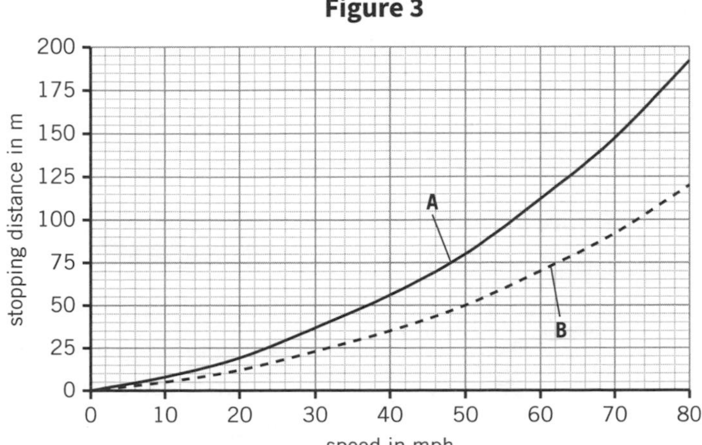

If you can't tell by looking at the lines, try turning each line in to a sentence. For example, when Car A was travelling at 80 mph, the stopping distance was 92.5 m.

09.4 The car is travelling at 50 mph (22.3 m/s). The reaction time of the driver is 0.4 s. Calculate the thinking distance for the car on each surface. Use the information in **Figure 3** to calculate the braking distance on both surfaces. **[6 marks]**

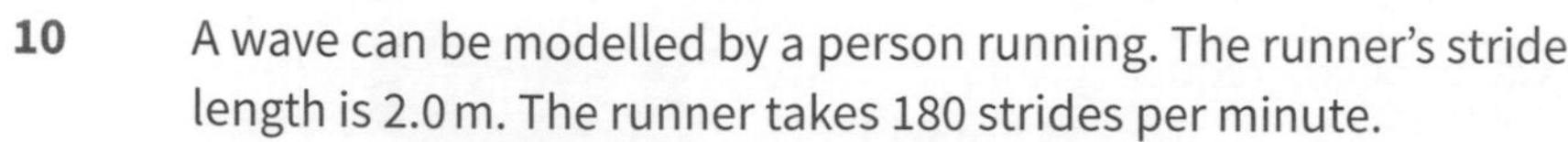

10 A wave can be modelled by a person running. The runner's stride length is 2.0 m. The runner takes 180 strides per minute.

10.1 Use the information to work out the speed of the runner in metres per second. **[3 marks]**

10.2 Write down which quantity is analogous to the wavelength. **[1 mark]**

10.3 Write down which quantity is analogous to frequency. Explain your answer. **[2 marks]**

10.4 Compare the method of working out the speed of a wave used in **10.1** with using the wave equation to work out the speed of a wave. **[3 marks]**

10.5 A student considers the limitations of the model. Explain **two** limitations of this model in terms of predicting and explaining what happens when a wave hits a boundary. **[3 marks]**

11 An oscilloscope can be used to display sound waves. The oscilloscope is connected to a microphone. The microphone converts sound waves into an alternating potential difference which is displayed on the screen (**Figure 4**).

Figure 4

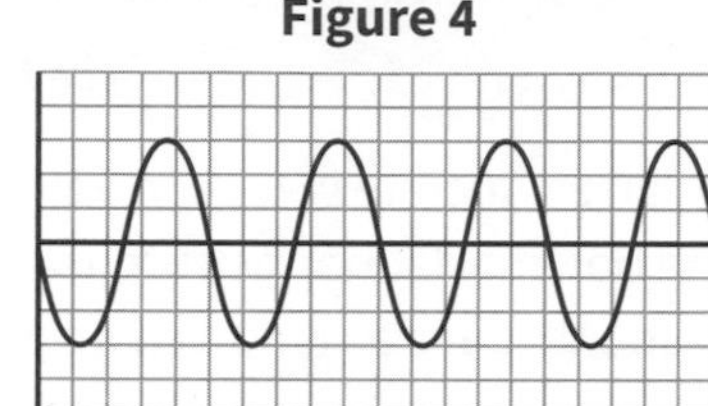

Exam Tip

ms is milliseconds, not to be confused with m/s, which is speed in meters per second.

11.1 Each square on the horizontal axis represents a time of 0.1 ms. Calculate the period of the sound waves. Give your answer in standard form. **[3 marks]**

11.2 Write down the equation that links wave speed, frequency, and wavelength. **[1 mark]**

11.3 The speed of sound waves in air is 340 m/s. Calculate the wavelength of the wave. **[3 marks]**

11.4 Each square on the vertical axis represents a potential difference of 2 V. Calculate the amplitude of the sound waves in volts. **[2 marks]**

Exam Tip

If it helps you can draw axis on **Figure 4** and label them with the values given to help you work out **11.4**.

12 A student wants to investigate the deflection of a beam. They clamp a ruler to the edge of a table, and attaches a mass to the end of the ruler as shown in **Figure 5**.

They decide to investigate how the length of the ruler affects the deflection.

Figure 5

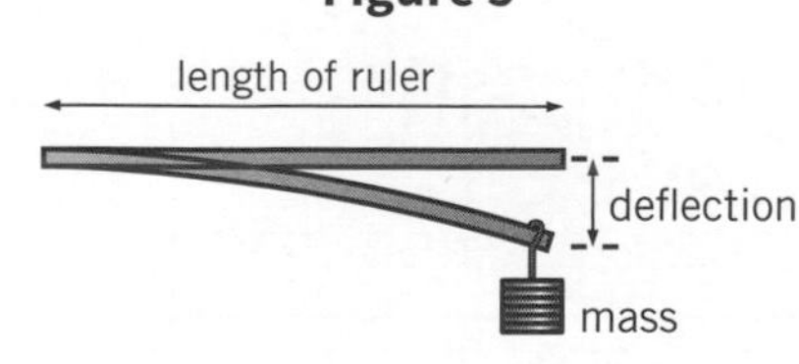

12.1 Write down the independent variable, the dependent variable, and two control variables. **[4 marks]**

Exam Tip

Remember:

- independent variable – what you change
- dependent variable – what you measure
- control variables – what you keep the same.

12.2 Describe a method that the student could use to collect data to find a relationship between the length of the ruler and the deflection. Assume that the equipment is set up as in **Figure 5**. **[3 marks]**

12.3 The student collects the data in **Table 1**.

Table 1

Length of ruler in m	Deflection in cm
0.20	3.5
0.30	3.8
0.40	4.2
0.50	5.3

The deflection is not proportional to the length. Use the data in **Table 1** to show this. **[2 marks]**

12.4 Suggest a different investigation that the students could carry out using the same equipment. **[1 mark]**

13 A baker uses their car to travel to work, and back home at the end of the day.

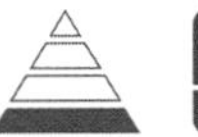

13.1 Write down the total displacement from the baker's home. **[1 mark]**

13.2 The distance from the baker's home to their workplace is 4.5 miles. There are 1609 m in one mile. Calculate 4.5 miles in metres. **[1 mark]**

13.3 The total time they spend in the car each day is 20 min. Calculate the average speed of her travel. **[4 marks]**

13.4 Describe why **13.3** is an average speed. **[1 mark]**

14 A car has broken down. A group of people decide to push the car to try to start it. For the car to start, the car needs to be travelling at 5 mph.

14.1 There are 1609 m in a mile. Show that 5 mph is approximately 2.2 m/s. **[3 marks]**

2.2 m/s is meters per second, whereas mph is miles per hour. In an hour you have 60 minutes each with 60 seconds.

14.2 Write down the equation that links acceleration, change in velocity, and time. **[1 mark]**

14.3 The group of people push the car with a constant acceleration. It takes 5.0 s to accelerate the car from 0 m/s to 2.2 m/s. Calculate the acceleration of the car. **[2 marks]**

14.4 Write down the equation that links force, mass, and acceleration. **[1 mark]**

14.5 Calculate the force exerted by the group of people. The mass of the car is = 1250 kg. **[3 marks]**

14.6 After the car has started it travels for a short distance at a constant speed of 5 mph before accelerating to 30 mph. Compare the magnitude of the driving and resistive forces: while travelling at 5 mph, while accelerating to 30 mph. You do not need to include any calculations. **[3 marks]**

Knowledge

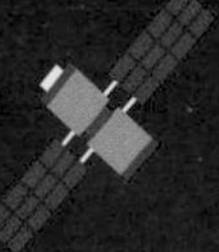
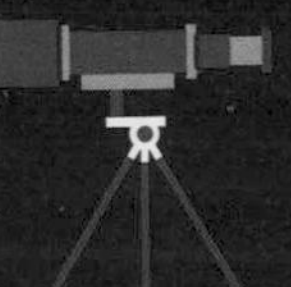

P15 Electromagnetic waves

The electromagnetic spectrum

Electromagnetic (EM) waves are **transverse** waves that transfer energy from their source to an absorber. For example, infrared waves emitted from a hot object transfer thermal energy.

Different substances may absorb, transmit, **refract**, or **reflect** EM waves in ways that vary with their wavelength.

Refraction occurs when there is a difference in the velocity of an EM wave in different substances.

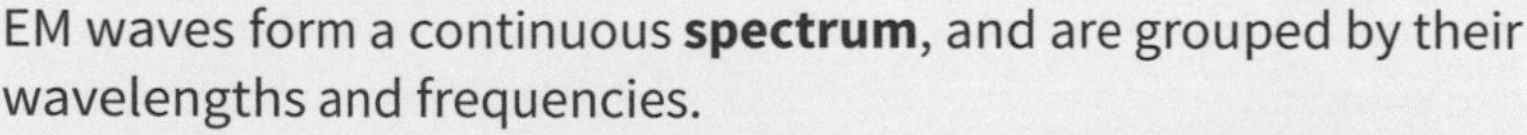

EM waves form a continuous **spectrum**, and are grouped by their wavelengths and frequencies.

EM waves all travel at the same velocity through air or a vacuum. They travel all at a speed of 3×10^8 m/s through a vacuum.

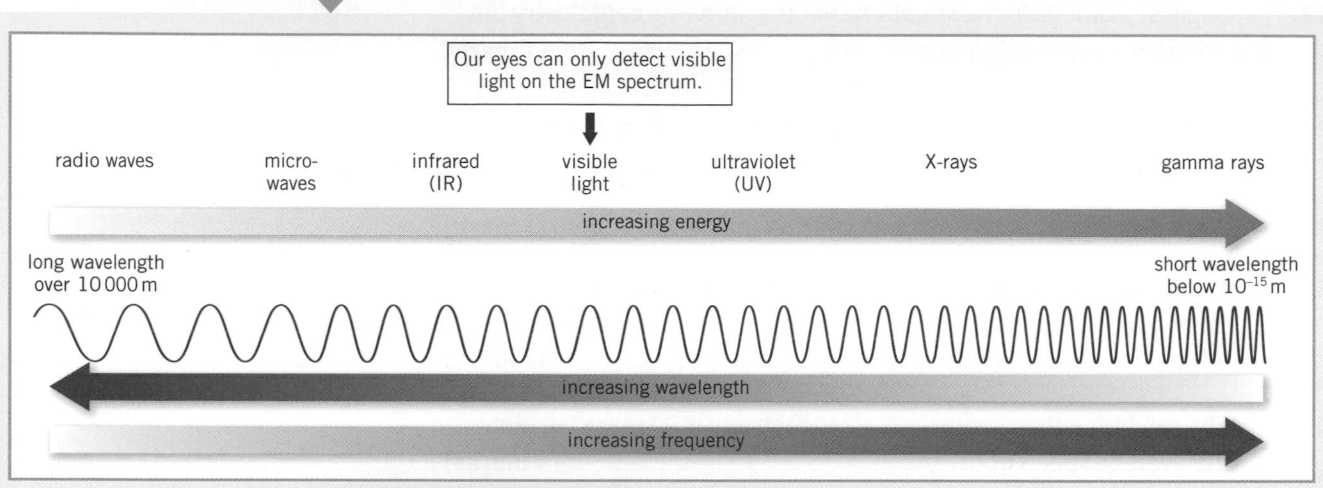

Refraction of electromagnetic waves

Ray diagrams show what happens when a wave is refracted (changes direction) at the boundary between two different substances.

- If a wave slows down when it crosses the boundary, the refracted ray will bend towards the normal.
- If a wave speeds up when it crosses the boundary, the refracted ray will bend away from the normal.
- If a wave travels at a right angle to the boundary (along the normal), it will change speed but not direction.

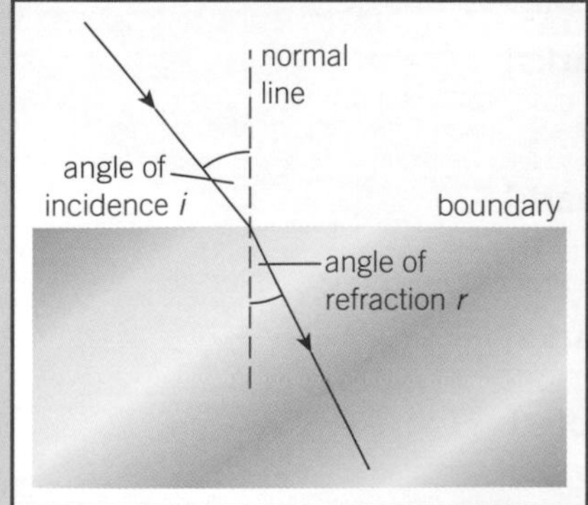

How to draw a ray diagram is covered in Chapter 14 *Mechanical waves*.

Wave front diagrams can be used to explain refraction in terms of the change of speed that occurs when a wave travels from one substance to another.

The wave front is an imaginary line at right angles to the direction the wave is moving.

- If a wave slows down as it crosses a boundary, the wave fronts become closer together.
- When a wave crosses a boundary at an angle, one end of the wave front changes speed before the other, so the wave changes direction.

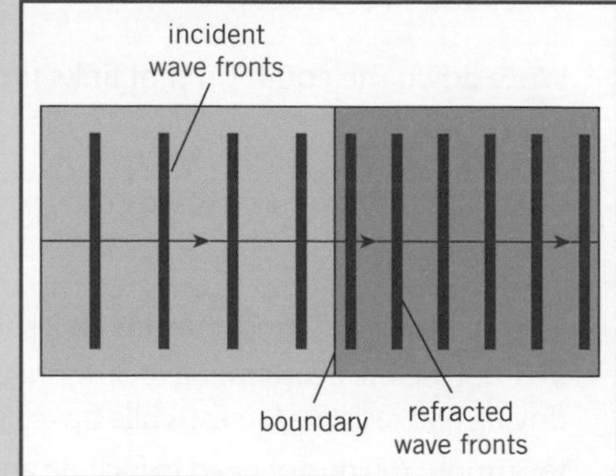

Properties of EM waves

EM waves of a wide range of frequencies can be absorbed or produced by changes inside an atom or nucleus. For example, gamma rays are produced by changes in the nucleus of an atom.

When electrons in an atom move down between energy levels, they emit EM waves.

Properties of radio waves

Radio waves can be produced by **oscillations** in an electrical circuit.

When radio waves are absorbed by a receiver aerial, they may create an **alternating current** with the same frequency as the radio waves.

Uses of EM waves

EM waves have many practical applications, but exposure to some EM waves (such as those that are forms of ionising radiation) can have hazardous effects.

Radiation dose (in sieverts) is the risk of harm from exposure of the body to a particular radiation.

Type of EM wave	Use	Why is it suitable for this use?	Hazards
radio waves	television and radio signals	• can travel long distances through air • longer wavelengths can bend around obstructions to allow detection of signals when not in line of sight	can penetrate the body and cause internal heating
microwaves	satellite communications and cooking food	• can pass through Earth's atmosphere to reach satellites • can penetrate into food and are absorbed by water molecules in food, heating it	
infrared	electrical heaters, cooking food, and infrared cameras	• all hot objects emit infrared waves – sensors can detect these to turn them into an image • can transfer energy quickly to heat rooms and food	can damage or kill skin cells due to heating
visible light	fibre optic communications	• short wavelength means visible light carries more information	can damage the retina
ultraviolet (UV)	energy efficient lights and artificial sun tanning	• carries more energy than visible light • some chemicals used inside light bulbs can absorb UV and emit visible light	can damage skin cells, causing skin to age prematurely and increasing the risk of skin cancer, and can cause blindness
X-rays	medical imaging and treatments	• pass easily through flesh, but not denser materials like bone • high doses kill living cells, so can be used to kill cancer cells – gamma rays can also be used to kill harmful bacteria	form of ionising radiation – can damage or kill cells, cause mutation of genes, and lead to cancers
gamma rays			

Key terms Make sure you can write a definition for these key terms.

alternating current · electromagnetic wave · electromagnetic spectrum · oscillation · radiation dose · ray diagram · reflection · refraction · transverse · wave front diagram

Retrieval

Learn the answers to the questions below then cover the answers column with a piece of paper and write as many as you can. Check and repeat.

	P15 questions	Answers
1	Are electromagnetic (EM) waves longitudinal or transverse waves?	transverse
2	Explain why EM waves are not mechanical waves.	they can travel through a vacuum (don't need a substance to travel through)
3	What do EM waves transfer from their source to an absorber?	energy
4	List the different types of waves in the EM spectrum in order of decreasing wavelength (increasing frequency).	radio, microwave, infrared, visible, ultraviolet, X-rays, gamma
5	Which part of the EM spectrum can humans see?	visible light
6	How can electromagnetic waves be produced?	changes inside an atom/atomic nucleus
7	How are gamma rays produced?	changes in the nucleus of an atom, for example during radioactive decay
8	How can radio waves be produced?	oscillations in an electrical circuit
9	How can we detect radio waves?	waves are absorbed and create an alternating current with the same frequency as the radio wave
10	What are radio waves used for?	transmitting television, mobile phone, and Bluetooth signals
11	What are microwaves used for?	satellite communications, cooking food
12	What is infrared radiation used for?	heating, remote controls, infrared cameras, cooking food
13	Which types of EM waves are harmful to the human body?	ultraviolet, X-rays, gamma rays
14	What are the hazards of being exposed to ultraviolet radiation?	damage skin cells, sunburn, increase risk of skin cancer, age skin prematurely, blindness
15	Why are X-rays used for medical imaging?	they pass through flesh but not bone
16	Why are gamma rays used for treating cancer and sterilising medical equipment?	high doses kill cells and bacteria
17	What is refraction?	waves change speed and direction as they cross the boundary from one substance to another due to the change in velocity
18	What happens to the direction of a refracted EM wave when it slows down as it crosses the boundary from one substance to another?	bends towards the normal

Put paper here

Now go back and use the questions below to check your knowledge from previous chapters.

Previous questions | Answers

	Previous questions	Answers
1	Give an example of a longitudinal wave.	sound waves
2	What can be done to reduce the drag on an object?	streamlining
3	Why does the pressure in a liquid increase with depth?	pressure at any point in a liquid is due to the weight of the liquid above that point
4	Give an example of a transverse wave.	electromagnetic waves
5	Is force a vector or scalar quantity?	vector
6	What is the name of the tendency of an object to remain in a steady state at rest or moving in a straight line at a constant speed?	Inertia

Put paper here

Put paper here

Required Practical

Investigate the amount of infrared radiated from different types of surfaces

Infrared radiation

This practical investigates the rates of absorption and radiation of infrared radiation from different surfaces.

You should be able to plan a method to determine the rate of cooling due to emission of infrared radiation, and evaluate your method.

To be accurate and precise in your investigation you need to:

- use an infrared detector with a suitable meter, where possible
- ensure that you always put the detector the same distance from the surface
- repeat measurements and calculate an average.

Worked example

A student wants to investigate the infrared radiation emitted by a surface coated with different colours of paint.

1 Describe a method to investigate the effect of surface colour on infrared emission rate.

Paint identical containers with different colours of paint. Fill each container with the same volume of water at the same temperature, and place on a heatproof mat. Place an infrared detector at the same distance and position relative to each container. Record the reading on the detector. Repeat three times.

2 Describe the type of graph the student should plot. Explain your answer.

A bar chart, because the surface colours are categoric data.

Practice

A student paints four jars with shiny and matt black paint, and shiny and matt white paint. They fill the jars with water at room temperature (20°C), and set the jars outside in the sun. After one hour they record the temperature increase of each jar.

Jar	A	B	C	D
Temperature increase in °C	5.5	8.0	7.0	1.5

1 Suggest the resolution of the instrument used to measure temperature.

2 Write down which jar (A, B, C, or D) is covered with matt black paint. Explain your answer.

3 Suggest two improvements the student could make to the experiment. Give an explanation for each improvement.

Practice

Exam-style questions

01 There are lamps that can be used to produce a suntan without needing to sunbathe.

The lamps emit ultraviolet radiation.

01.1 Name an electromagnetic wave with longer wavelength than ultraviolet radiation. **[1 mark]**

____

01.2 Name an electromagnetic wave with higher frequency than ultraviolet radiation. **[1 mark]**

____

01.3 Describe one hazard of using an ultraviolet lamp. **[1 mark]**

____

____

01.4 People who work outside are exposed to ultraviolet radiation from the Sun.

Suggest one method of reducing the risk of injury. **[1 mark]**

____

____

02 A teacher demonstrates how different materials absorb electromagnetic radiation.

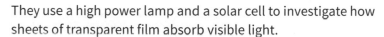

They use a high power lamp and a solar cell to investigate how sheets of transparent film absorb visible light.

They fix the lamp 10 cm from the solar cell, and connect the solar cell to a voltmeter. The solar cell produces a potential difference that is proportional to the intensity of the light that falls on it.

Table 1 shows the data that they obtained.

Table 1

Number of sheets	Potential difference in V			
	Repeat 1	Repeat 2	Repeat 3	Mean
0	5.87	5.05	6.12	5.68
1	4.12	3.61	4.55	4.10
2	3.74	6.75	4.40	
3	3.12	2.81	3.13	3.02
4	2.89	2.02	3.86	2.92
5	2.67	2.64	3.21	2.84

Go through the text and pick out the key points before attempting the questions. You may want to highlight the important information.

02.1 Explain why the potential difference decreases as the number of sheets of transparent film increases. **[2 marks]**

____

____

____

____

02.2 Calculate the missing mean in **Table 1**. **[1 mark]**

Mean: ______________

> **Exam Tip**
> Whenever you're asked to calculate a mean, always check for any outliers.

02.3 Suggest how a source of systematic error might arise in the experiment, and how it could be reduced. **[2 marks]**

____

____

____

> **Exam Tip**
> The marks for this question are for giving an error *and* how to reduce it. Do not list two errors, you will only gain 1 mark. Don't give two errors and describe how one can be reduced either. You won't gain an extra mark, and if the extra error is wrong you may lose a mark! Only ever give exactly what is asked in the question.

02.4 The teacher repeats the experiment with a gamma radiation source.

Suggest why the teacher ensures that all students are standing at least 1 m from the gamma source. **[2 marks]**

____

____

____

02.5 Compare the way that visible light is produced by the lamp with the way that gamma rays are produced by the source.

Use ideas about atoms, electrons, and nuclei in your answer. **[2 marks]**

____

____

____

____

> **Exam Tip**
> The question gives you a big clue as to what the examiners are looking for in an answer – ideas about atoms, electrons, and nuclei. Make sure you talk about these in your answer.

03 Radiographers use X-ray machines in hospitals for medical imaging, for example computerised tomography (CT) scans.

A narrow X-ray beam circles around one part of your body. It produces lots of 'slices' of the body which the computer puts together to make a detailed image of an organ, bone or blood vessel.

Table 2 shows the radiation doses from different types of CT scan.

Table 2

Examination	Average effective dose in mSv
head	2
spine	6
chest	15

03.1 A radiographer must leave the room when a CT scan is being taken.

Explain why they should leave the room. **[2 marks]**

____

____

> **Exam Tip**
> Radiographers do several scans in one day, nearly every day of the week.

03.2 Define the term radiation dose. **[1 mark]**

____

____

03.3 A traditional X-ray of a foot produces an effective radiation dose of 0.001 mSv.

Suggest why the dose is much smaller than a CT scan of the head. **[1 mark]**

____

____

03.4 Calculate the number of foot X-rays that would give a patient the same dose as a CT scan of the head. **[2 marks]**

> **Exam Tip**
> We've said it before and we'll say it again. Always show your working! that way, if you make a mistake or get the wrong answer, you can still gain the marks.

number = ____________

03.5 We also receive radiation from natural sources. This is called background radiation. In the UK, the average annual dose is 2.7 mSv.

Calculate the number of years of background radiation that is equivalent to a CT scan of the chest. **[2 marks]**

____

____

____

04 A student has made a list of appliances in his house that use different electromagnetic waves.

He identifies the television and the electrical heater as two appliances that use electromagnetic waves. However, he cannot find appliances for some of the other waves in the electromagnetic spectrum.

Explain how televisions and heaters use electromagnetic waves.

Suggest uses for **two** of the waves for which the student cannot find appliances. **[6 marks]**

> **Exam Tip**
>
> There are lots of parts to this question. First, identify the electromagnetic radiation that can be used in the home. Then work out which electromagnetic waves are left over and think of uses for them. The answer needs to have *two* different sources of electromagnetic waves included.

05 The atmosphere only allows certain wavelengths of the electromagnetic spectrum to reach the Earth's surface.

05.1 Write down the electromagnetic wave that our eyes detect. **[1 mark]**

05.2 Radio waves also reach the Earth's surface. People sometimes confuse radio waves and sound waves.

Complete the sentences with words from the box below. The words can only be used once. **[5 marks]**

> **Exam Tip**
>
> Cross off each word as you use it, this way you won't get confused and use a word twice.

transverse	340 m/s	matter	air	300 000 km/s	longitudinal	energy

Waves of the electromagnetic spectrum are ____________ waves that travel at ____________.

Sound waves are ____________ waves and travel at ____________.

All waves transfer ____________ without transferring ____________.

05.3 Some microwaves reach the surface of the Earth from the Sun.

Write down which wave, radio or microwave, has the lower frequency. **[1 mark]**

> **Exam Tip**
>
> Learning the order of the electromagnetic waves is really important. There are lots of fun mnemonics to help your remember, you just need to find one you like and can remember!

05.4 We sometimes use the power of microwaves to cook food.

Write down a use of microwaves other than cooking. **[1 mark]**

06 **Figure 1** shows the change in direction of waves as they move between two different media.

Figure 1

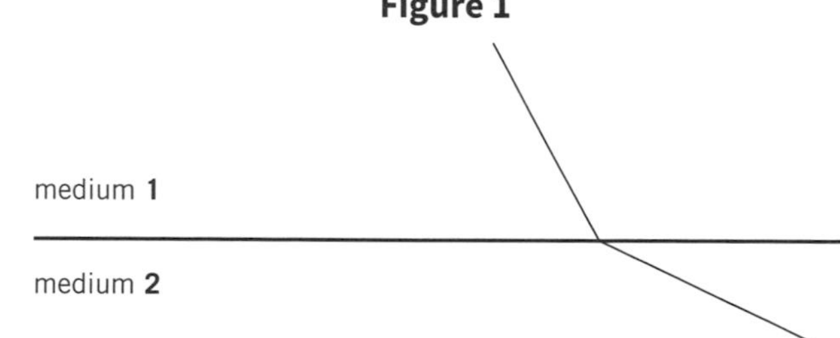

06.1 Add **three** lines to show the wavefronts of waves moving in medium **1**. **[2 marks]**

06.2 Add **three** lines to show the wavefronts of waves moving in medium **2**. **[3 marks]**

> **Exam Tip**
>
> Whether the waves bend towards or away from the normal depends on whether the waves speed up or slow down as they pass into the medium 2. This is the first thing you have to work out.

06.3 Describe what happens to the speed and frequency of the waves when they move from medium **1** to medium **2**. **[2 marks]**

06.4 Suggest a situation where waves would behave as shown in the diagram. **[1 mark]**

07 Electromagnetic waves have many different uses. We use visible light to communicate.

07.1 Name **two** other electromagnetic waves that are used for communication. In each case describe the use. **[2 marks]**

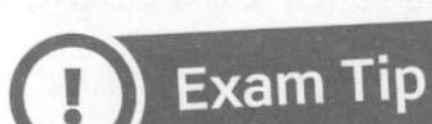

The question gives you visible light as an example, so you won't get any marks for that in your answer to question **07.1**.

07.2 Some waves, such as gamma rays, can penetrate the human body. Describe a use of another electromagnetic wave that can penetrate the body. **[2 marks]**

07.3 Compare the changes in the atom that produce gamma rays with the changes in the atom that produce visible light. **[2 marks]**

08 A driver in a car sees an obstacle ahead on the road. The motion of the car is shown in **Figure 2**.

Figure 2

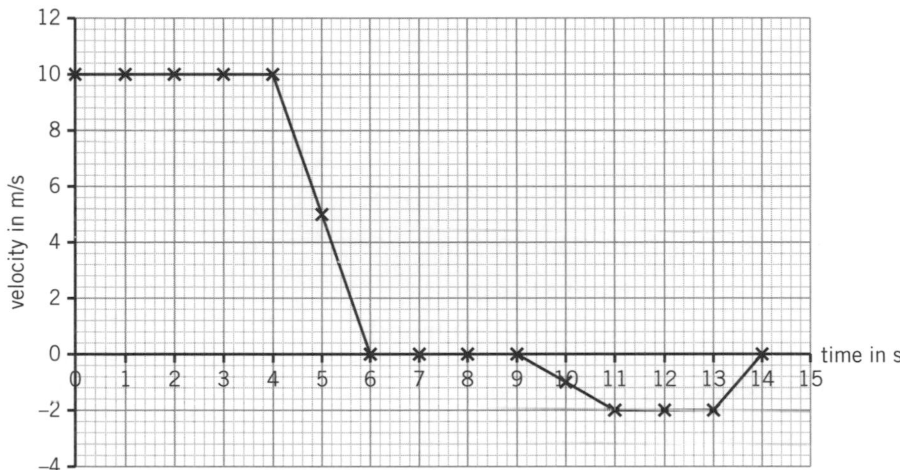

08.1 Identify the time at which the driver saw the obstacle. Give a reason for your answer. **[2 marks]**

08.2 Calculate the distance travelled by the car when it was travelling at its initial constant speed. **[2 marks]**

08.3 Calculate the acceleration of the car between 4 and 6 seconds. Comment on whether the answer is positive or negative. **[3 marks]**

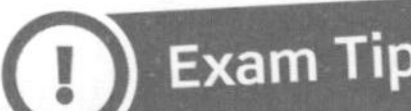

Draw lines on the graph to help with your working.

08.4 Compare the acceleration of the car between 9 and 11 seconds with the acceleration that you calculated in **08.3**. **[2 marks]**

09 In 1867 James Clark Maxwell developed a model that linked electricity and magnetism. Maxwell predicted that it was possible to produce electromagnetic waves even though no one had ever detected them.

09.1 Describe one difference between a model and a description. **[2 marks]**

09.2 Radio waves were first produced and detected by Heinrich Hertz in 1887. Explain why it was important for Hertz to produce and detect the waves. **[1 mark]**

09.3 Today we can use an alternating current to produce radio waves. Radio waves can be detected using an aerial. An aerial is attached to a device, such as a radio, that needs an alternating current to work. Describe what happens inside the aerial when the radio wave is absorbed. **[3 marks]**

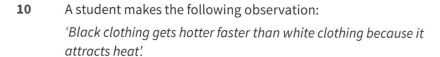

> **Exam Tip**
>
> This is a two mark question. You'll need to say how it refers to a model and then use words like 'whereas' or 'however' to contrast a model to a description.

10 A student makes the following observation:

'Black clothing gets hotter faster than white clothing because it attracts heat.'

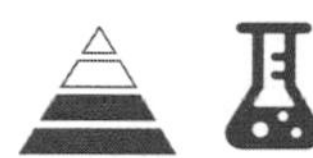

10.1 Re-write the sentence so that it is correct. **[1 mark]**

10.2 The student makes a second observation:

'Silver surfaces only reflect radiation. They do not absorb or emit radiation.'

Suggest a situation where we use the fact that silver surfaces reflect radiation. **[1 mark]**

10.3 A student sets up a silver can and a black can. They fill the cans with hot water, put a lid on each can, and record the temperature of the water using a datalogger.

Describe **two** variables that they need to control in order to make a comparison between the cans. **[2 marks]**

10.4 Sketch the graph of temperature against time that the student would obtain from their experiment. **[4 marks]**

10.5 Suggest whether the experiment supports the students second observation in **10.2**. **[2 marks]**

> **Exam Tip**
>
> The number of marks give you important clues as to what you need to do in a question. **10.1** is worth 1 mark so you only need to make a small change.

11 A gardener wants to purchase a greenhouse. There is a choice of different types of glass to put in the greenhouse.

Figure 3 shows the percentage of sunlight that is transmitted through each type of glass.

Figure 3

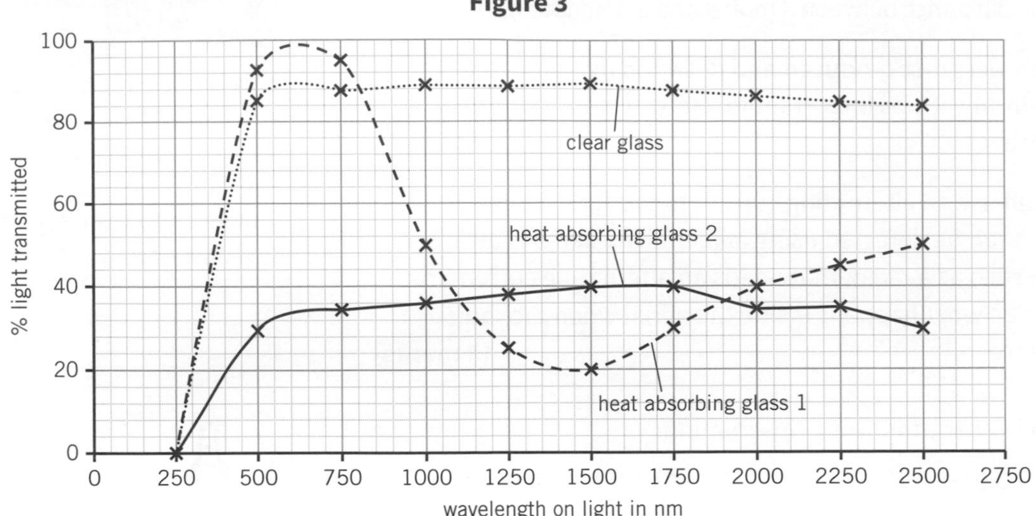

11.1 Ultraviolet radiation has a wavelength from 250 nm to 400 nm. Infrared radiation has a wavelength above 750 nm.

Use this information to write down the range of wavelengths of visible light. **[1 mark]**

> **Exam Tip**
>
> Don't let the very small numbers put you off. Just treat them the same as you would any other number.

11.2 Write down which type of glass is the best at absorbing radiation with a wavelength of 2500 nm.

Justify your answer. **[2 marks]**

11.3 Plants need to absorb radiation in the range 430 nm and 662 nm.

Suggest which type of glass would be best for growing plants.

Justify your answer. **[2 marks]**

11.4 Suggest whether you could get a suntan inside a greenhouse made using heat absorbing glass 2.

Justify your answer. **[3 marks]**

> **Exam Tip**
>
> You need to use data to justify your answer to **11.4**.

12 A student reads that the wavelength of X-rays is about the same as the diameter of an atom. She recalls that the radius of an atom is 1×10^{-10} m. The speed of electromagnetic radiation is 300 000 km/s.

12.1 Calculate the frequency of X-rays. **[5 marks]**

12.2 Write down the order of magnitude (power of 10) of the frequency of X-rays. **[1 mark]**

12.3 The student learns that the wavelength of microwaves is about 10 cm. Show that the order of magnitude of the frequency of microwaves is 10^9 Hz. **[3 marks]**

> **Exam Tip**
>
> Be careful with the unit for frequency. You must write the unit correctly, with a capital H and a lowercase z. Anything else and you won't gain the mark.

13 A car and a boat are travelling at 35 mph (16 m/s). The car uses its brakes and comes to a stop in a distance of 30 m. The boat comes to a stop in a distance of 100 m after the engine is turned off. Both the car and the boat have a mass of 1500 kg.

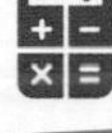

Exam Tip

List all the important information from the question.

car:

- initial speed = 16 m/s
- mass = 1500 kg
- stopping distance = 30 m

boat:

- initial speed = 16 m/s
- mass = 1500 kg
- stopping distance = 100 m

13.1 Identify the force acting on the car and the force acting on the boat that brings them to a stop. **[2 marks]**

13.2 Draw a free body diagram for the boat for when its engine has been turned off and it is moving through the water but has not yet come to a stop.

Assume the boat is moving from left to right. **[3 marks]**

13.3 Calculate the kinetic energy of the boat. **[2 marks]**

13.4 The force acting on the boat brings it to a stop over a certain distance. This means the force does work on the boat.

Use the equation that links work, force, and distance to calculate the force required to stop the boat.

Suggest whether the braking force acting on the car is bigger, smaller, or the same size as the force acting on the boat.

Give reasons for your answer. **[6 marks]**

Exam Tip

Not all the information you need will be in one location. You'll need to use the data from the main body of the question to answer this.

14 A student accelerates a toy car. They use a newtonmeter to apply a constant force. They use light gates to measure the speed at two different times.

14.1 Write down the equation that links change in velocity, time, and acceleration. **[1 mark]**

14.2 The initial velocity is 0.5 m/s. The final velocity is 2.7 m/s. The time between the measurements of velocity is 0.4 s.

Calculate the acceleration of the trolley. **[2 marks]**

Exam Tip

Even if it doesn't specifically ask you to, get into the habit of writing down the equation you are using as your first step whenever you do a calculation. You'll can be given a mark for writing the correct equation.

14.3 Write down the equation that links mass, force, and acceleration. **[1 mark]**

14.4 The student applies a force of 2.0 N. The mass of the trolley is 400 g (0.4 kg).

Calculate the acceleration of the trolley. **[3 marks]**

14.5 Suggest one practical reason why the two values of acceleration are not the same. **[1 mark]**

Knowledge

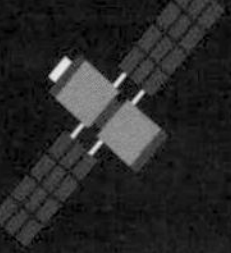

P16 Light and sound

Sound waves

Sound waves are:

mechanical waves – they need a solid, liquid, or gas medium to travel through (cannot travel through a vacuum)

longitudinal – the oscillations of particles in the medium are *parallel* to the direction of energy transfer.

When sound waves go from air into a solid, they cause vibrations of the same frequency in the solid.

Hearing

Sound waves cause solid parts of the ear (e.g., ear drum) to vibrate.

The brain converts these vibrations into what we hear, but only over a limited frequency range (20 Hz to 20 kHz).

higher frequency = higher pitch

greater amplitude = louder sound

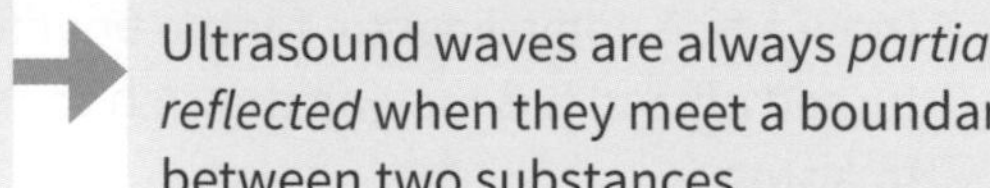

Uses of sound waves

Ultrasound waves have a frequency above the range of human hearing.

Ultrasound waves are always *partially reflected* when they meet a boundary between two substances.

The distance to a boundary can be found by timing how long it takes for an ultrasound reflection to come back to a detector. This can be used for medical and industrial imaging.

An echo is a reflected sound wave.

Echo sounding uses high-frequency sound waves to detect objects in deep water and to measure the depth of water.

Seismic waves

The centre of the Earth cannot be studied directly but the way seismic waves travel through the Earth can provide evidence about the internal structure of the Earth.

Seismic waves' paths curve, showing us that there is a *gradual change in the density of the mantle.*

S-waves are not detected on the opposite side of the Earth, suggesting *the outer core must be liquid.*

Kinks in the paths of P-waves travelling to the other side of the Earth suggest a sudden change in the density of the material they are passing through.

The existence of shadow zones, where no P-waves are detected, suggest the existence of a *solid inner core.*

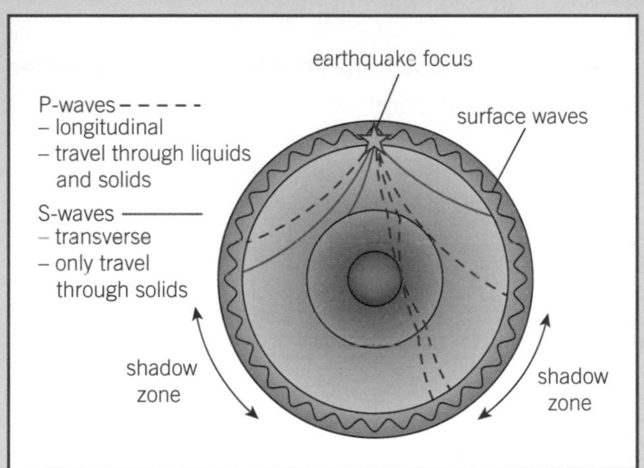

Revision tips

Ray diagrams may look confusing, but drawing them step-by-step can make them logical and simple.

Learn the rules to draw them rather than memorising the diagrams themselves.

Key terms

Make sure you can write a definition for these key terms.

concave, converge, convex, diffuse, diverge, focal length, longitudinal, opaque, P-wave, real image, S-wave, specular, translucent, transparent, transverse, ultrasound, virtual image

Convex lenses

Convex lenses curve outwards. They make parallel rays of light **converge** at a point. **Focal length** is the distance from the centre of the lens to the principal focus.

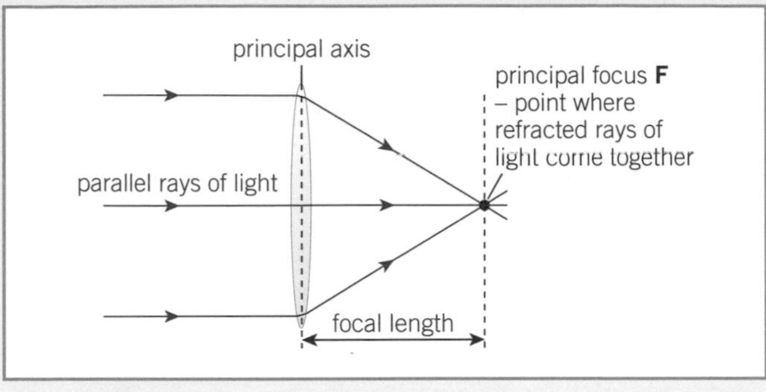

Forming images

Images formed by **convex** lenses can be either real or virtual.

Real images can be projected onto a screen. **Virtual images** appear to come from behind the lens.

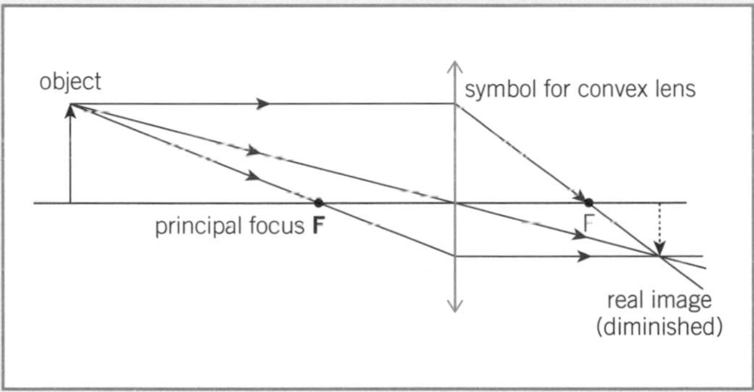

Concave lenses

Concave lenses curve inwards. They make parallel rays of light **diverge** (so they appear to come from a point).

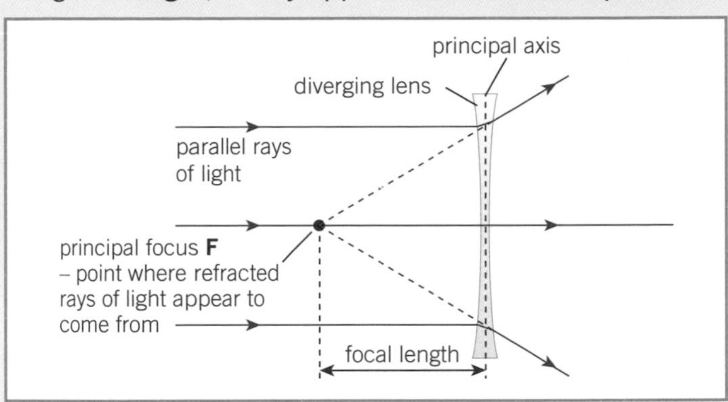

Forming images

Images formed by **concave** lenses are always virtual.

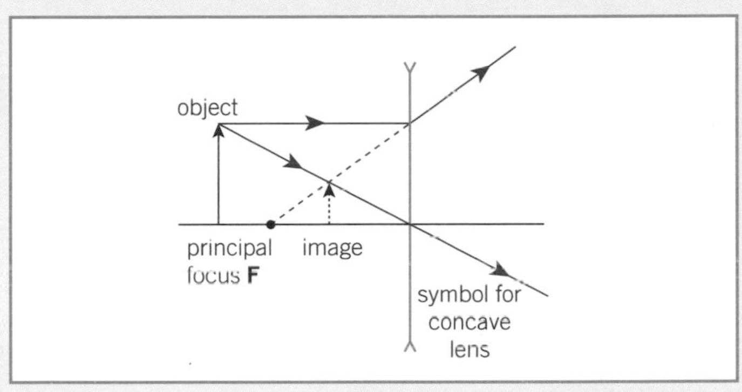

Images formed by a lens can be:

- magnified of diminished
- upright or upside down (inverted).

The magnification of an image can be calculated using:

$$\text{magnification} = \frac{\text{image height}}{\text{object height}}$$

Magnification has no units because it is a ratio.

Visible light

Each colour within the visible light spectrum has its own narrow band of wavelength and frequency.

Reflection from:

- a smooth surface in a single direction is called **specular** reflection
- a rough surface causes *scattering* of light (**diffuse** reflection).

Transparent objects transmit visible light.

Translucent objects transmit visible light, but light rays are scattered or refracted inside them.

Opaque objects do not transmit visible light, but absorb and reflect it.

The colour of an object depends on the wavelengths they transmit and reflect.

Coloured filters work by absorbing certain wavelengths of light and transmitting others.

Retrieval

Learn the answers to the questions below then cover the answers column with a piece of paper and write as many as you can. Check and repeat.

	P16 questions	Answers
1	Are sound waves transverse or longitudinal?	longitudinal
2	Why are sound waves mechanical?	need a substance to travel through
3	What do sound waves do to the ear drum and other solid parts of the human ear?	make them vibrate
4	Why is there a limit to the range of human hearing?	conversion of sound waves in air to vibrations in ear (solid) only works over a limited range of frequencies
5	What is the frequency range of normal human hearing?	20 Hz to 20 000 Hz (20 kHz)
6	What are ultrasound waves?	sound waves with a frequency above 20 000 Hz
7	What happens to ultrasound waves when they meet a boundary between two substances?	they are partially reflected and some are transmitted
8	What is echo sounding used for?	detecting objects in deep water, measuring the depth of water
9	What are the properties of P-waves?	longitudinal, travel through liquids and solids
10	What are the properties of S-waves?	transverse, cannot travel through liquids
11	Which observation suggests that the outer core of the Earth must be liquid?	S-waves are not detected on the opposite side of the Earth
12	What is the difference between a concave and convex lens?	convex bulges out in the middle, concave is thinner in the middle than at the edges
13	What does a convex lens do to parallel rays of light?	light converges (comes together) at the principal focus
14	What does a concave lens do to parallel rays of light?	light diverges (spreads out) so they appear to have come from the principal focus
15	What is the focal length of a lens?	distance from the centre of the lens to the principal focus
16	What kind of images do concave and convex lenses produce?	concave = virtual, convex = real or virtual
17	What properties do all EM waves of the same colour share?	same range of wavelengths and frequencies
18	What four things can happen to visible light when it hits an object?	transmitted, absorbed, reflected, or refracted
19	What is the difference between specular and diffuse reflection?	specular = reflection from smooth surface, diffuse = reflection from rough surface
20	What words describe an object that transmits visible light?	transparent or translucent
21	Why does an object appear opaque?	does not transmit visible light – absorbs and reflects it
22	How do colour filters work?	absorb certain wavelengths of light, transmit others

Put paper here

Now go back and use the questions below to check your knowledge from previous chapters.

Previous questions

Answers

	Question	Answer
1	What are radio waves used for?	transmitting television, mobile phone, and Bluetooth signals
2	What is a transverse wave?	oscillations/vibrations are perpendicular (at right angles) to the direction of energy transfer
3	What are three factors that can affect the braking distance of a vehicle?	speed, road conditions, condition of tyres and brakes
4	What is the resultant force on an object moving at a steady speed in a straight line?	zero
5	What is the difference between distance and displacement?	distance is a scalar quantity and only has a magnitude (size), displacement is a vector quantity and has both magnitude and direction
6	What will an object placed in a fluid do if its weight is equal to the upthrust?	float
7	What is inelastic deformation?	when an object does not go back to its original shape and size when deforming forces are removed

Put paper here

Put paper here

Put paper here

Required Practical

Reflection and refraction

In this practical, you should have traced rays of light from a ray box as they interact with different surfaces or materials.

This includes investigating how light refracts as it passes through different materials, and how light is reflected by different surfaces.

To carry out accurate and precise investigations you need to:

- use low light conditions
- place the slit in the ray box as far from the bulb as possible
- use a sharp pencil and ruler to draw the rays
- draw a line at 90° to any surface or boundary and measure all angles from this line to the ray
- mark either side of solid block to work out the path of a ray inside the block.

Worked example

A student wants to determine the angle of reflection for a particular angle of incidence.

They set up a ray box and a mirror, and marked on paper the paths of the rays.

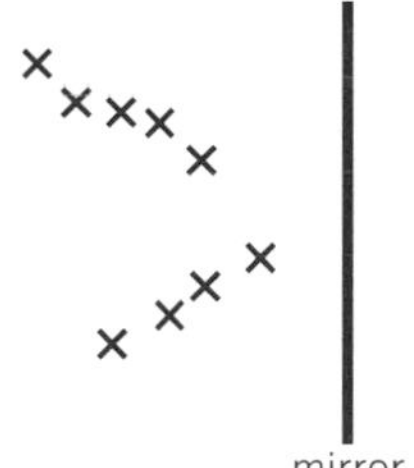

1 Explain how you know the student has drawn the mirror in the wrong place.

The rays will not meet on the line where the mirror is placed.

2 Suggest one reason why the marks for each ray are not in a straight line.

The beam from the ray box was very broad because the slit was too close to the bulb, or the classroom lights were on.

Practice

A student sets up a ray box and directs the ray at a glass block. They produce the drawing shown.

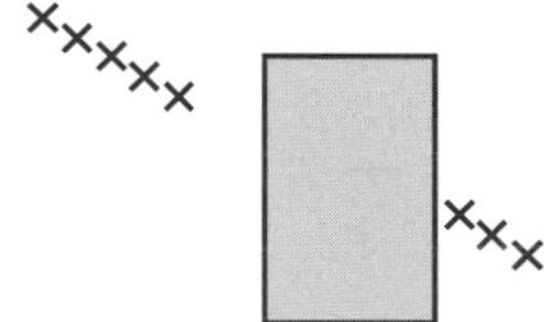

1 Complete the diagram by drawing in the rays using the marks that the student put on the paper.

2 Label the angle of incidence, angle of refraction, and the normal.

3 Explain why the rays entering and leaving the block are parallel.

Practice

Exam-style questions

01 A student is looking through convex and concave lenses.

01.1 Complete the sentences. Use the words in the box. **[4 marks]**

concave	convex

The student sees only a virtual image when they look through a ______________ lens. The image through a ______________ lens can be real or virtual.

A ______________ lens spreads light rays out. A ______________ lens brings light rays to a focus.

> **Exam Tip**
>
> Each 'fill in the gaps' question is different. For this one you have to use word given more than once. In other ones, you can only use each word one time. Always read the question carefully.

01.2 The student notices that an image is magnified.

The image height is 1.2 cm. The object height is 0.7 cm.

Calculate the magnification.

Use an equation from the *Physics Equations Sheet*. **[2 marks]**

Magnification = ______________

01.3 Suggest why magnification has no unit. **[1 mark]**

____

____

01.4 Draw a diagram to show how light from a distant object is refracted by a concave lens. **[3 marks]**

> **Exam Tip**
>
> Use a ruler and a pencil for **01.4**. It requires accuracy and you won't get that if you draw it free hand.

02 A tsunami warning centre has a seismometer.

The seismometer detects earthquake waves, also called seismic waves.

A strong earthquake on the other side of the Earth produces P-waves and S-waves.

02.1 Compare P-waves and S-waves in terms of:

- whether they are transverse or longitudinal
- what they can travel through. **[4 marks]**

____

____

____

____

____

Exam Tips

02.1 asks you *two* things about *two* different waves, so it gives you the structure of your answer:

P-waves are transverse / longitudinal and can travel through…

S-Waves are transverse / longitudinal and can travel through…

02.2 There is an earthquake on one side of the Earth.

The seismometer, on the opposite side of the Earth, detects a P-wave after 26 minutes.

The radius of the Earth is 6400 km.

Calculate the average speed of the waves in km/s.

Explain why it is an average speed. **[5 marks]**

Exam Tip

Think about the structure of the Earth, and the different states – solid, liquid, or gas – each part of the Earth is in.

Speed = ______________ km/s

Explanation: ____

____

02.3 The seismometer does not record any S-waves.

Explain why. **[2 marks]**

____

____

02.4 Another seismometer, closer to the origin of the earthquake, detects a P-wave that has travelled through the crust of the Earth.

The second seismometer also detects the wave after 26 minutes.

Compare the speed of the P-wave in the crust with the speed of the wave travelling through the Earth. **[2 marks]**

____

____

____

02.5 Seismic waves travel faster in denser rock.

Suggest which rock has the higher density. **[1 mark]**

____

____

03 A student buys a dog whistle. When she blows on the whistle, she cannot hear the sound but her dog can hear it.

03.1 Suggest a frequency for the sound that the whistle is producing. **[1 mark]**

03.2 The student can hear the dog when it barks. Describe the difference between the sound wave produced by the dog when it barks, and the sound wave produced by the whistle. **[1 mark]**

> **Exam Tip**
>
> For a compare question you need to give ways in which the bark and the whistle have both *similar* affects and *different* affects on the ear drum.

03.3 Describe the effect of the sound waves on the ear drum. Compare the effect of the sounds of the bark and the whistle on the ear drum. Use your answer to explain why you cannot hear the whistle. **[4 marks]**

03.4 Suggest a device that converts the disturbance in a sound wave to the vibration of a solid. **[1 mark]**

03.5 The student and the dog are in a park. There are buildings nearby. She notices that she can hear the echo of the bark as it is reflected off the buildings. Suggest whether the reflection is specular or diffuse. Give reasons for your answer. **[2 marks]**

04 A student wears a red shirt.

04.1 Explain why the shirt appears to be red. **[2 marks]**

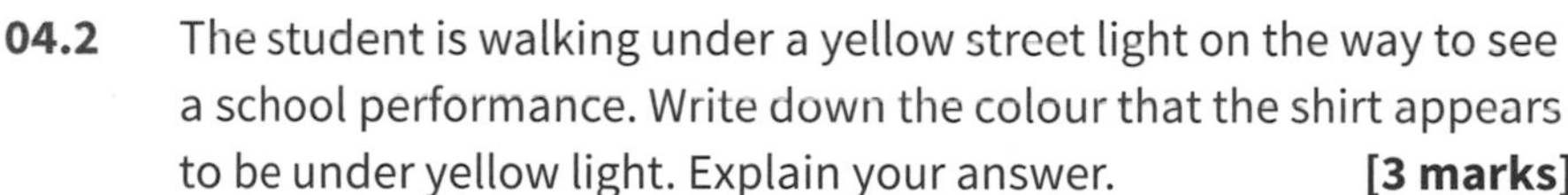

04.2 The student is walking under a yellow street light on the way to see a school performance. Write down the colour that the shirt appears to be under yellow light. Explain your answer. **[3 marks]**

> **Exam Tip**
>
> An explain question means you give to give a detailed why.

04.3 At the performance, the student looks up and sees that some of the stage lights have coloured filters in front of them. One of the filters produces a circle of green light on the stage. Describe what the green filter does to white light. **[2 marks]**

04.4 An actor with a blue shirt stands in the circle of green light. Suggest why the actor's blue shirt does not look completely black. **[2 marks]**

05 Very little was known about the interior of the Earth until the analysis of seismic waves.

05.1 Explain why scientists need to use a model when describing the inside of the Earth. Write down the name of another scientific model that is used for a similar reason. **[3 marks]**

> **Exam Tip**
>
> There are lots of examples you could pick for **05.1**, but the one you are probably most familiar with is very very small.

05.2 In 1692, Edmond Halley proposed what is now known as the Hollow Earth model. His model described the Earth as a hollow shell that was about 500 miles thick, with an inner sphere and air in between. Suggest one reason why the Hollow Earth model was not adopted. **[1 mark]**

05.3 In the early twentieth century, scientists observed that there were areas on the Earth's surface where no earthquake waves were detected. These are called shadow zones. They are shown in **Figure 1**.

On a diagram of the Earth, draw lines to show the paths of P and S waves that travel through the Earth. Explain why the regions that P and S waves reach are different. Write down the feature of the current model of the Earth that shadow zones do **not** suggest.

[5 marks]

Figure 1

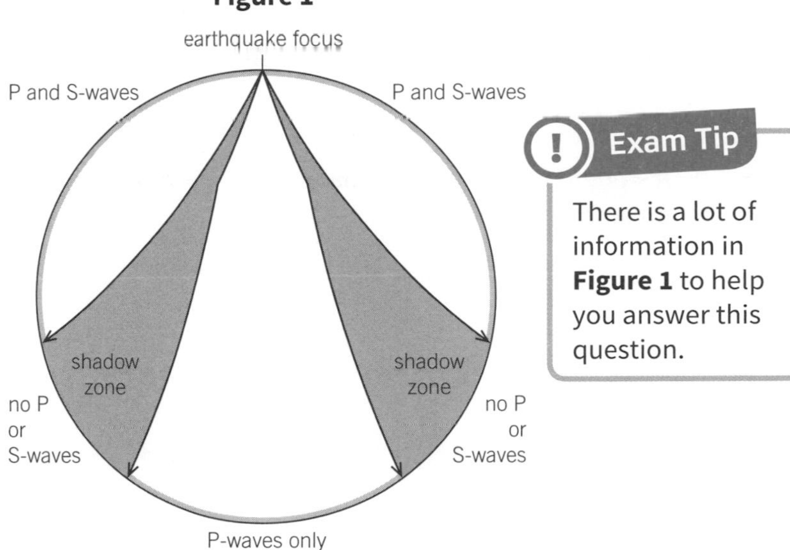

> **Exam Tip**
> There is a lot of information in **Figure 1** to help you answer this question.

05.4 In 2018, a group of scientists published a paper that disputed the current ideas about the how the inner core of the Earth was formed. Suggest the process that occurred before the paper was published.

[1 mark]

06 A student is confused about colour. His science teacher has explained that when you mix all the colours together you produce white. In art lessons the student uses paint to change the colour that you see. Paint is an opaque material. His art teacher has told him that when you mix all the colours together you get black. Describe what is being 'mixed' in science lessons compared to art lessons. Explain the difference between what is observed using ideas of reflection and absorption. **[6 marks]**

> **Exam Tip**
> The key to **06** is the information that paint is opaque.

07 A student looks through a thick perspex rod at some text on a piece of paper. They notices that when the rod is close to the text, the image of the text is magnified. However, when the rod is further from the text, the image is diminished. The rod is behaving like a lens.

> **Exam Tip**
> The steps for drawing ray diagrams are the same no matter where the object is. Just follow the rules and you should see the differing results.

07.1 Write down which type of lens shows the same behaviour. **[1 mark]**

07.2 Complete the diagrams in **Figure 2** to show how the image changes as the student moves the rod away from the text. The focal points are shown as 'f'. **[6 marks]**

Figure 2

Text (T) closer to rod:

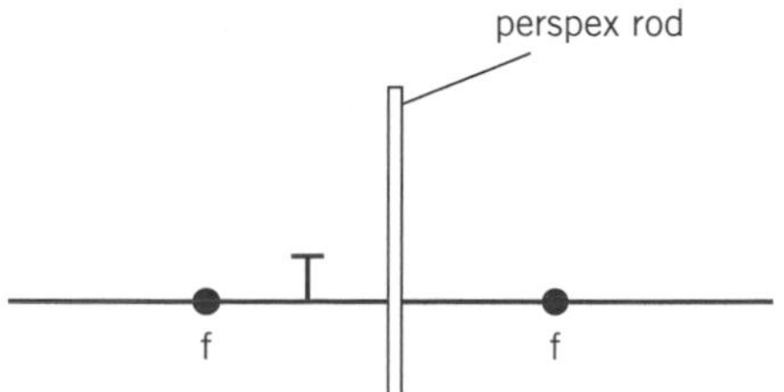

Text (T) further from rod:

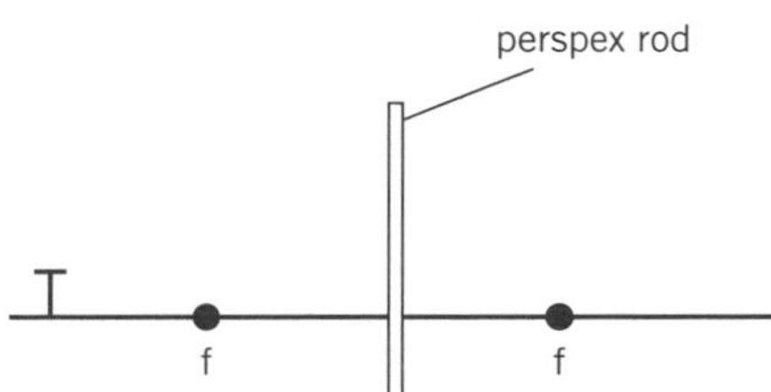

07.3 The student makes another observation about the change to the image as she moves the rod away. Use the diagrams to describe the second observation about the image. **[2 marks]**

08 A student finds a website that shows how much they would weigh on different planets.

08.1 Describe one difference between the weight and the mass of an object. **[2 marks]**

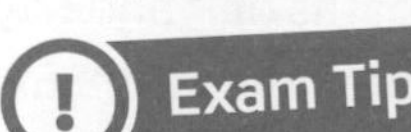

In a calculation you can get a mark for substituting in the correct values and a mark for rearranging the equation. Make sure you clearly demonstrate both steps in your working.

08.2 Write down the equation that links mass, weight, and gravitational field strength. **[1 mark]**

08.3 The website says that the student's weight on Mercury would be 179 N. The gravitational field strength on Mercury is 3.8 N/kg. Calculate the mass of the student. **[3 marks]**

08.4 The student sits on a sofa. The sofa contains springs that compress when the student sits down. Describe the relationship between the weight of the student and the compression of the springs. **[2 marks]**

09 A skateboarder is traveling at 2.0 m/s. The mass of the skateboarder is 54 kg.

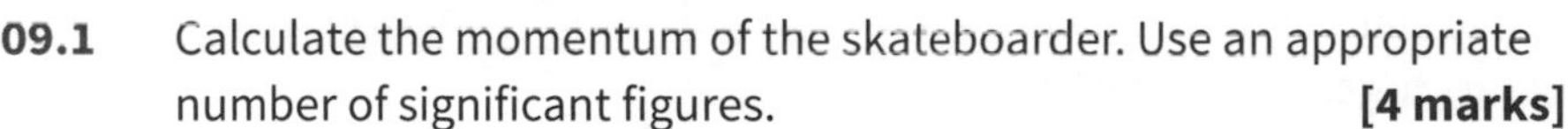

09.1 Calculate the momentum of the skateboarder. Use an appropriate number of significant figures. **[4 marks]**

09.2 As the skateboarder moves along, they grab their backpack from the ground and continue moving forward. When no other forces act, their velocity decreases. Explain why. **[2 marks]**

09.3 The skateboarder comes to a stop using friction between their shoe and the ground. The frictional force is 80 N. Calculate the acceleration of the skateboarder. **[3 marks]**

10 A boat uses ultrasound to find the depth of a lake.

10.1 Define ultrasound. **[1 mark]**

10.2 Write down the equation that links speed, distance, and time. **[1 mark]**

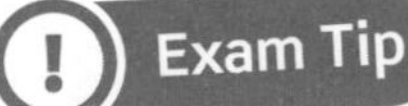

Remember the sound wave has to go to the bottom of the lake and back before it is detected.

10.3 The speed of sound in water is 1500 m/s. The boat sends out a pulse of ultrasound. The pulse is detected after 16 ms. Calculate the depth of the lake. **[5 marks]**

10.4 A student uses an ultrasonic emitter and receiver to investigate how to use sound to measure distance. She measures the time for the beam to be reflected from an object and calculates the distance. She then measures the actual distance with a ruler and writes her data in **Table 1**.

Table 1

Time for reflection in ms	Calculated distance in m	Actual distance in m
9	1.53	1.51
6	1.02	1.00
3	0.51	0.49

The student thinks that there could be an error in the ultrasonic distance measurement. Use the data to suggest the type of error. Explain your answer. **[2 marks]**

11 An technician in a hospital uses an ultrasonic transducer to produce an image of a foetus. An ultrasonic transducer emits a series of pulses of ultrasound. The transducer also contains a receiver that can convert a reflected pulse into a potential difference. The path of one of the pulses is shown in **Figure 3A**.

Figure 3

A

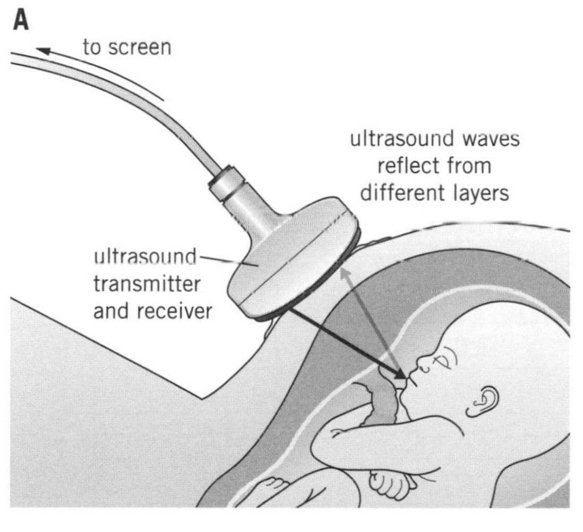

B

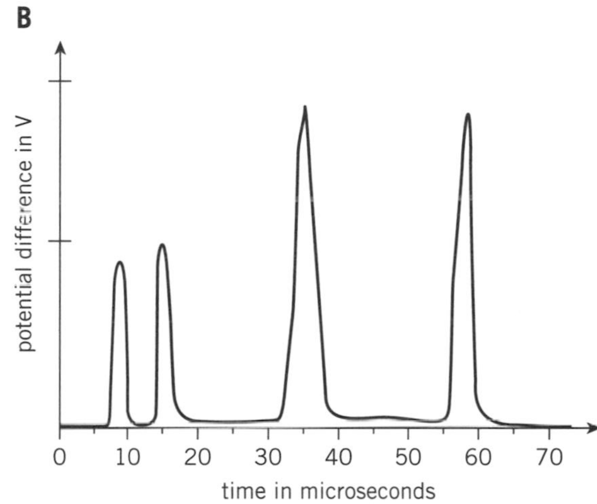

11.1 Suggest why there are four peaks in **Figure 3B**. Explain your answer. **[2 marks]**

11.2 The fetus' mouth produces the third pulse. Calculate the distance between the bottom of the transducer and the fetus. Assume that the speed of sound in the tissue is 1540 m/s. **[5 marks]**

11.3 The resolution of the transducer is 1 mm. Define resolution. **[1 mark]**

11.4 The resolution of the transducer is approximately the same as the wavelength of the ultrasound. Write down the equation that links the wavelength, speed, and frequency of a wave. **[1 mark]**

11.5 Calculate the frequency of the ultrasound. Write your answer in standard form. **[5 marks]**

12 A student investigates the magnification of an object by a lens. He sets up a lamp, a lens, and a screen as shown in **Figure 4**. He sees an image of the filament of the lamp on the screen.

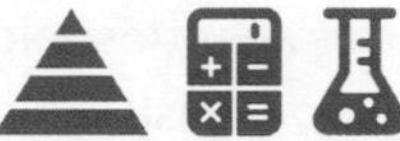

Figure 4

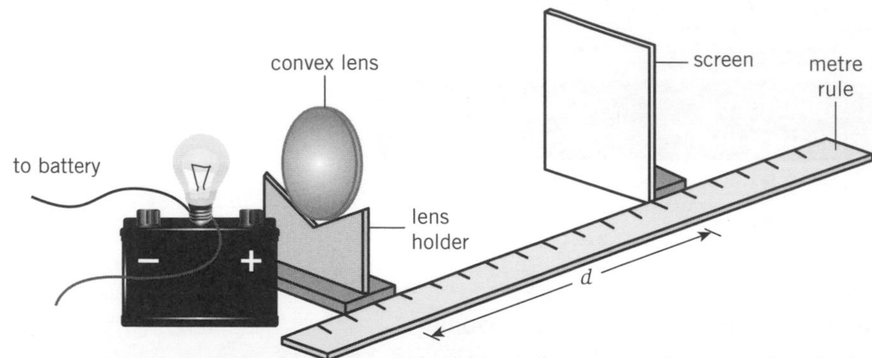

The student uses a ruler to measure the distance d between the lens and the screen and the size of the image on the screen.

12.1 Estimate the uncertainty in measurement when using a ruler to measure distance. **[1 mark]**

Exam Tip

Look at the divisions on the ruler.

12.2 As the student moves the lamp away from the lens the image becomes more blurred. Suggest why. **[1 mark]**

12.3 Describe what the student needs to do to produce an image in focus, with the lens and lamp in their new positions. Give reasons for your answer. **[2 marks]**

12.4 **Table 2** shows the student's data.

Table 2

d in cm	Image size in cm			
	Measurement 1	Measurement 2	Measurement 3	Mean
12.1	2.9	3.1	3.0	3.0
12.3	3.4	3.6	3.7	3.6
12.5	3.9	3.6	3.8	3.8
13.0	4.8	5.0	4.7	4.8

Describe how to use the range of measurements of image size to calculate the uncertainty in the measurement. **[1 mark]**

12.5 Calculate the uncertainty in the image size at d = 13.0 cm. Compare this uncertainty with the uncertainty in **12.1**. **[2 marks]**

12.6 The width of the filament of the lamp is 4 mm. Calculate the magnification of the lens when d = 13.0 cm. Use an equation from the *Physics Equations Sheet*. **[3 marks]**

13 A student watches as some potatoes are cooking in a saucepan. The saucepan has a lid, but the lid occasionally lifts up.

13.1 Describe in terms of forces why the lid moves up. **[1 mark]**

13.2 Explain in terms of particles, momentum, and Newton's laws why the steam exerts a force on the lid. **[3 marks]**

13.3 The student wants to calculate the pressure of the steam inside the pan needed to lift the lid. She writes a list of quantities that she knows:

- mass of pan lid = 300 g
- area of pan lid = 0.13 m^2
- gravitational field strength = 9.8 N/kg

Write down the equation that links mass, weight, and gravitational field strength. **[1 mark]**

13.4 Use the equation linking pressure, force, and area to calculate the minimum pressure of the steam needed to lift the lid. **[5 marks]**

13.5 Explain why the value is a minimum. **[2 marks]**

Exam Tip

You'll need to use two equations to solve this. Sometimes a question in an exam will link two equations together. You need to work with each one logically.

14 A student has downloaded an app onto his phone. The phone contains a device that can measure velocity as a function of time. While on a rollercoaster ride, the student records the graph shown in **Figure 5**.

Exam Tip

Treat each change in the graph as a different section with a different behaviour happening. You may find it helpful to highlight the line in different colours every time there is a change in the graph.

Figure 5

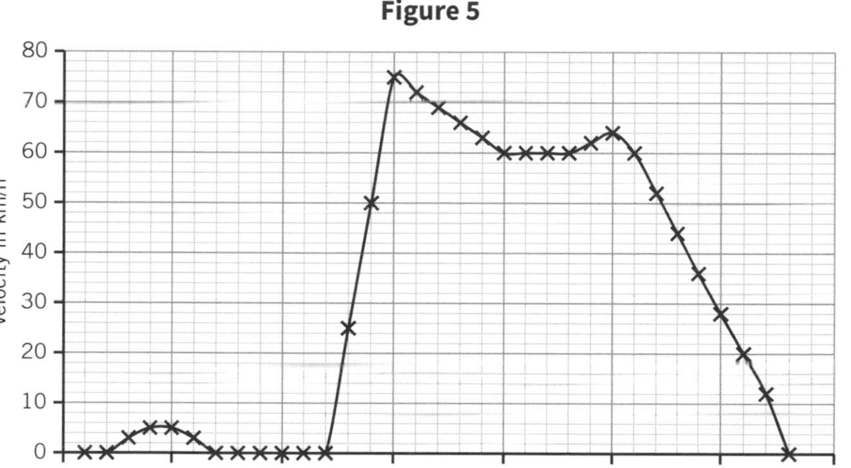

14.1 Write down a time interval during which the acceleration was large and positive. **[1 mark]**

14.2 Write down a time interval during which the acceleration was large and negative. **[1 mark]**

14.3 Write down what the student was doing between 7 seconds and 12 seconds. **[1 mark]**

14.4 Write down the time interval, after 13 seconds, where the net force on the student was zero. Explain your answer. **[2 marks]**

14.5 Use the information on **Figure 5** to estimate the distance travelled on the ride between 12 seconds and 33 seconds. Give your answer to an appropriate number of significant figures. **[5 marks]**

14.6 Suggest why you cannot use the equation 'distance = speed × time' to do this calculation. **[1 mark]**

Exam Tip

Divide the graph up into different sections, and then work out the area of each shape individually.

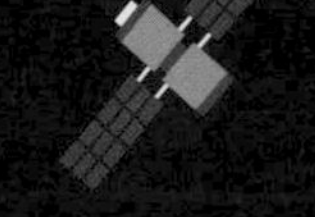
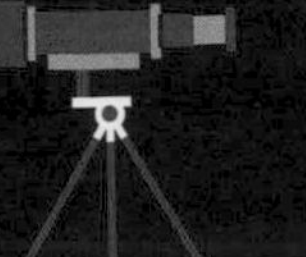

Knowledge

P17 Magnets and electromagnets A

Magnets

Magnets have a north (N) and a south (S) pole.

When two magnets are brought close together, they exert a non-contact force on each other.

Repulsion – If the poles are the same (N and N or S and S), they will repel each other.

Attraction – If the poles are different (N and S or S and N), they will attract each other.

The force between a magnet and a magnetic material (iron, steel, cobalt, or nickel) is always attractive.

Magnetic fields

A **magnetic field** is the region around a magnet where another magnet or magnetic material will experience a force due to the magnet.

A magnetic field can be represented by magnetic field lines.

Field lines show the direction of the force that would act on a north pole at that point.

Field lines always point from the north pole of a magnet to its south pole.

A magnetic field's strength is greatest at the poles and decreases as distance from the magnet increases.

The closer together the field lines are, the stronger the field.

Revision tip

The lines on a diagram that show magnetic field lines will always point from north to south.

Induced and permanent magnets

A **permanent** magnet produces its own magnetic field which is always there.

An **induced** magnet is an object that becomes magnetic when it is placed in a magnetic field.

The force between an induced magnet and a permanent magnet is *always attractive* (it doesn't matter which pole of the permanent magnet the induced magnet is near).

If the induced magnet is removed from the magnetic field it will quickly lose most or all of its magnetism.

Plotting magnetic fields

A magnetic compass contains a small bar magnet that will line up with magnetic field lines pointing from north to south.

A compass can be used to plot the magnetic field around a magnet or an **electromagnet**:

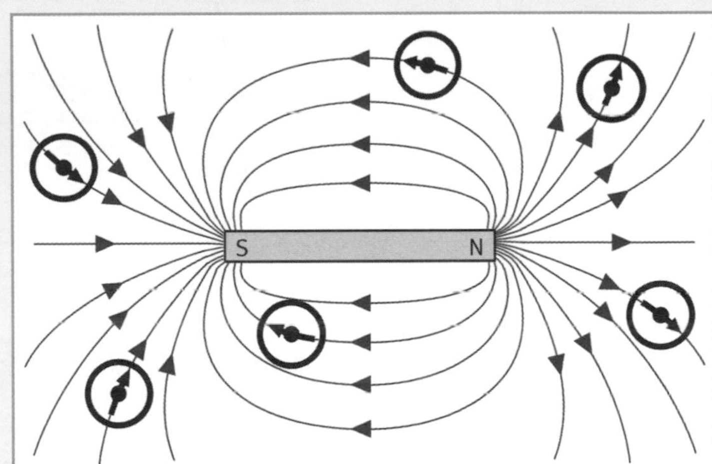

If it is not near a magnet, a compass will line up with the Earth's magnetic field, providing evidence that the Earth's core is magnetic.

As a compass points towards a south pole, the magnetic pole near the Earth's geographic North Pole is actually a south pole.

Key terms Make sure you can write a definition for these key terms.

attraction electromagnet induced magnetic field

Electromagnetism

If an electric current flows through a wire (or other conductor), it will produce a magnetic field around the wire.

The field strength increases:

- with greater current
- closer to the wire.

Reversing the direction of the current reverses the direction of the field.

The field around a straight wire takes the shape of concentric circles at right angles to the wire:

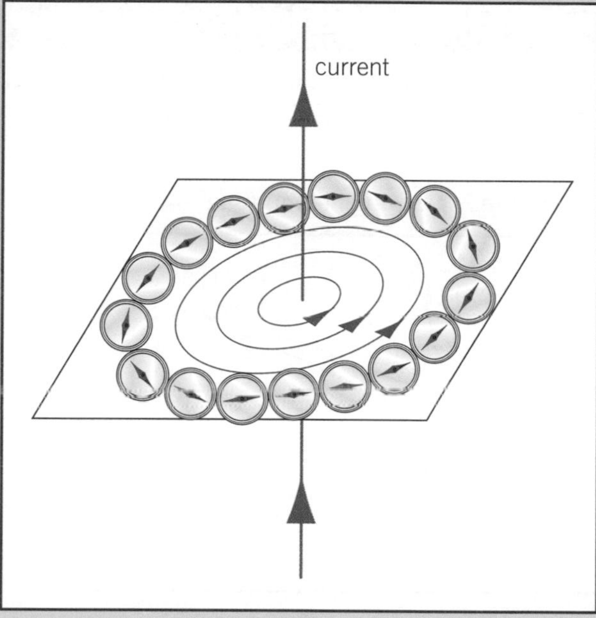

If the wire was gripped by someone's right hand so that the thumb pointed in the direction of the current, the fingers would curl in the direction of the magnetic field.

Solenoids

A **solenoid** is a cylindrical coil of wire.

Bending a current-carrying wire into a solenoid increases the strength of the magnetic field produced.

The shape of the magnetic field around a solenoid is similar to a magnetic field around a bar magnet.

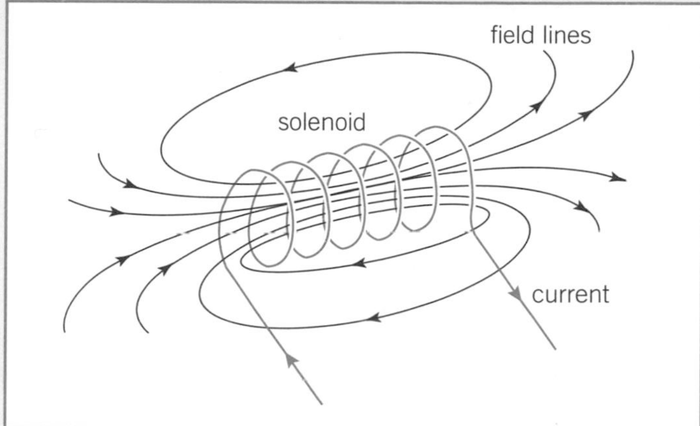

Inside a solenoid the magnetic field is *strong* and *uniform*, which means it has the same strength and direction at all points.

The strength of the magnetic field around a solenoid can be increased by putting an iron core inside it.

If the wire was gripped by someone's right hand so that the fingers curl in the direction of the current in the coil, the thumb will point towards the north pole of the field.

Electromagnets are often solenoids with an iron core.

Advantages of electromagnets

- An electromagnet can be turned on and off.
- The strength of an electromagnet can be increased or decreased by adjusting the current.

permanent repulsion solenoid

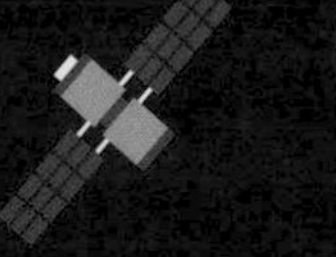

P17 Magnets and electromagnets B

Uses of electromagnets

Scrap-yard crane

Heavy objects containing magnetic materials can be lifted using an electromagnet.

Electric bell

The diagram below shows how an electric bell operates.

- switch is pressed, turning the electromagnet on
- the iron armature is attracted towards the electromagnet, making the hammer strike the gong
- the circuit is broken so the electromagnet stops working and the armature springs back
- circuit is complete again and the cycle starts again, continuing as long as the switch is pressed.

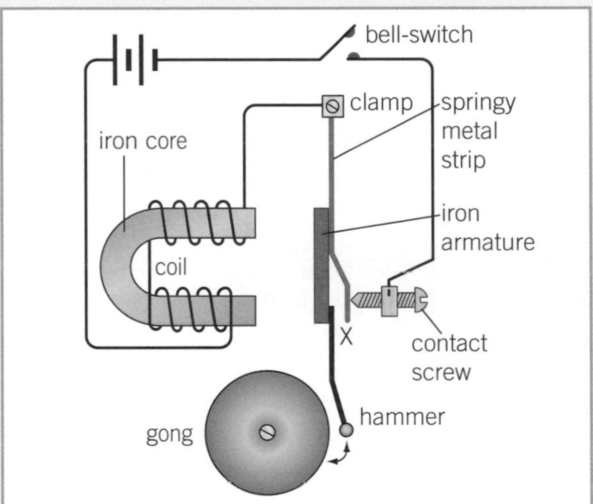

Circuit breaker

A switch that is in series with an electromagnet.

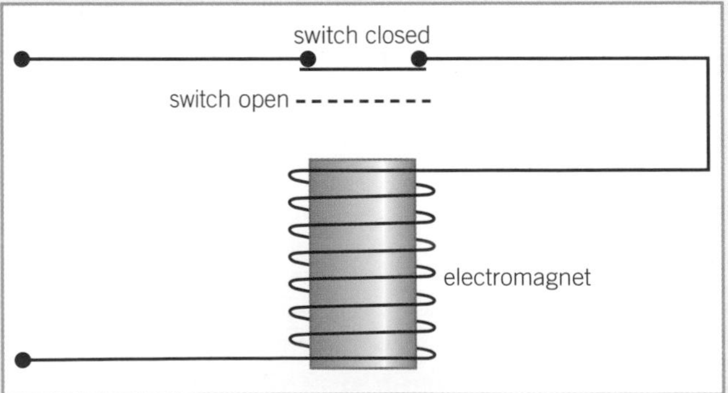

The switch is held closed by a spring, but if the current becomes too large, the electromagnet becomes strong enough to pull the switch into the open position, turning the current off.

The motor effect

When a current-carrying wire (or other conductor) is placed in a magnetic field, it experiences a force.

The force is due to the interaction between the field created by the current in the wire and the magnetic field in which the wire is placed.

The magnet producing the field will experience an equal-sized force in the opposite direction.

The direction of the force is reversed if the current is reversed or if the direction of the magnetic field is reversed.

Fleming's left-hand rule

The direction of the force/motion of the wire is always at right angles to both the current and the direction of the magnetic field it is within.

It can be worked out using Fleming's left-hand rule:

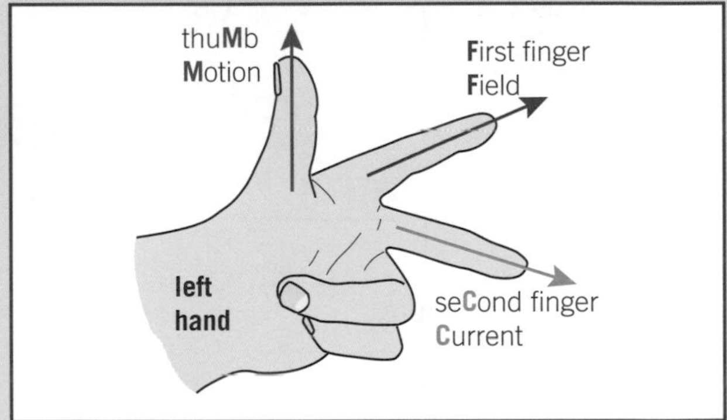

Magnetic flux density

The **magnetic flux density** of a field is a measure of the strength of the magnetic field.

For a current-carrying wire at right angles to a magnetic field, the size of the force on it is given by the equation:

force (N) = magnetic flux density (T) × current (A) × length (m)

$$F = BIl$$

Electric motors

A current-carrying coil of wire in a magnetic field will tend to rotate.

This is the basis of an electric motor.

The diagram below shows a simple motor made of one rectangular piece of wire.

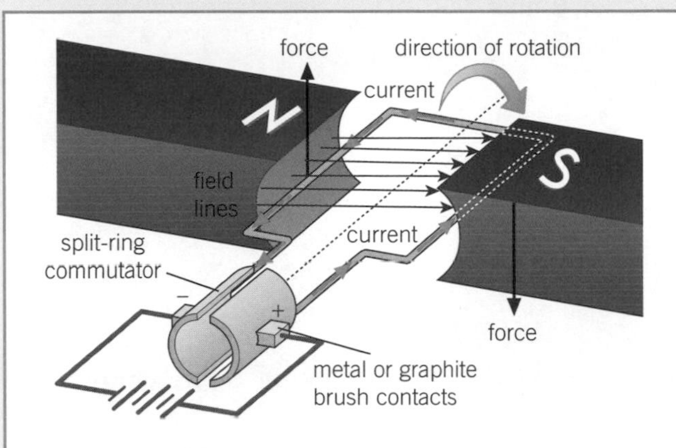

When there is a current in the wire, it spins because:

- each side of the coil experiences a force due to being a current-carrying conductor in a magnetic field
- the forces on each side of the coil are in opposite directions.

The **split-ring commutator** keeps the motor spinning in the same direction.

The ends of the wire swap contacts with the power supply every half turn, so current always flows in the same direction relative to the magnetic field.

The motor can be made to spin

- *faster* – by increasing the current in the coil or increasing the strength of the magnetic field.
- *in the opposite direction* – by reversing the direction of the current or reversing the direction of the magnetic field.

Loudspeakers

Moving-coil loudspeakers and headphones use the **motor effect** to convert changes of current in a coil of wire to changes of pressure in sound waves.

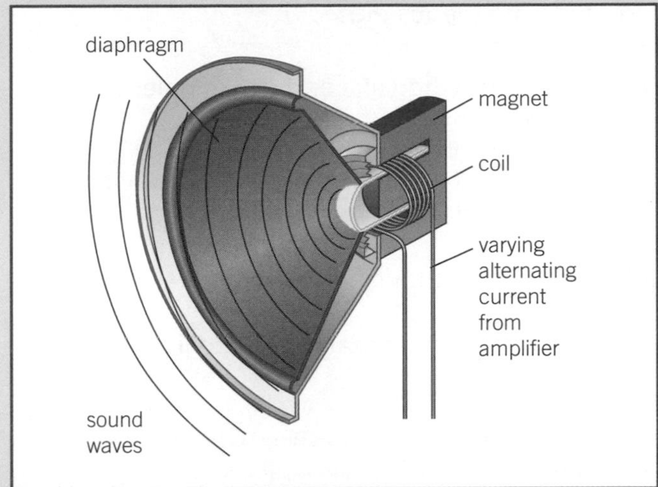

A coil of wire is placed inside a permanent magnet (so it is inside a magnetic field) and is attached to a diaphragm.

When a current flows through the coil, it experiences a force due to the motor effect.

This causes the diaphragm to move.

When the current changes direction, the force on the coil also changes direction, causing the diaphragm to move in the opposite direction.

Variations in the current make the coil and diaphragm vibrate.

These vibrations create variations of pressure in the air which form a sound wave.

The frequency of the sound wave produced is the same as the frequency of the alternating current supplied to the coil.

Key terms Make sure you can write a definition for these key terms.

magnetic flux density | motor effect | split-ring commutator

Retrieval

Learn the answers to the questions below then cover the answers column with a piece of paper and write as many as you can. Check and repeat.

	P17 questions	Answers
1	What is a magnetic field?	the region of space around a magnet where a magnetic material will experience a force
2	What happens when like and unlike poles are brought together?	like = repel, unlike = attract
3	What happens to the strength of the magnetic field as you get further away from the magnet?	decreases
4	Where is the magnetic field of a magnet strongest?	at the poles
5	In which direction do magnetic field lines always point?	north to south
6	What does the distance between magnetic field lines indicate?	strength of the field, closer together = stronger field
7	What is a permanent magnet?	material that produces its own magnetic field
8	What is an induced magnet?	material that becomes magnetic when it is put in a magnetic field
9	What does a magnetic compass contain?	small bar magnet
10	What is produced around a wire when an electric current flows through it?	magnetic field
11	What factors does the strength of the magnetic field around a straight wire depend upon?	size of current, distance from wire
12	What effect does shaping the wire into a solenoid have on the magnetic field strength?	increases strength of magnetic field
13	How can the strength of the magnetic field inside a solenoid be increased?	put an iron core inside
14	What does Fleming's left-hand rule show?	relative orientation of the force, current in the conductor, and magnetic field for the motor effect
15	What is the symbol for magnetic flux density and what unit is it measured in?	B, tesla (T)
16	What is the motor effect?	when a conductor placed in a magnetic field experiences a force
17	What causes the motor effect?	interaction between the magnetic field created by current in a wire and the magnetic field in which the wire is placed
18	What do loudspeakers and headphones do?	use the motor effect to convert variations in current in electrical circuits to pressure variations in sound waves

Put paper here

Now go back and use the questions below to check your knowledge from previous chapters.

Previous questions / Answers

	Previous questions	Answers
1	Are sound waves transverse or longitudinal?	longitudinal
2	What is the frequency range of normal human hearing?	20 Hz to 20 000 Hz (20 kHz)
3	What is a compression?	area in longitudinal waves where the particles are squashed closer together
4	What is a longitudinal wave?	oscillations/vibrations are parallel to the direction of energy transfer
5	What are the properties of S-waves?	transverse, cannot travel through liquids

Put paper here

Put paper here

Maths Skills

Plotting graphs

When plotting a graph, draw it so that it covers at least half the graph paper.

Use a sensible scale that makes it easy to plot and read the graph. Generally, each large square on the graph paper should represent a numerical value of 1, 2, 5, or 10.

The labels on each graph axis should give the name and unit of the variable plotted.

A line of best fit is a smooth line that passes through *or near* each plotted point – this can be curved or straight.

A linear graph is any straight line graph – if this goes through the origin the two variables are directly proportional.

The equation of all straight lines is:

$y = mx + c$

Where:

y = variable plotted on y-axis

m = gradient of line

x = variable plotted on x-axis

c = y-intercept

Worked example

A student measured the potential difference across a fixed resistor while varying the current.

Current in A	Potential difference in V
0.05	1
0.10	2
0.15	3
0.20	4
0.25	5
0.30	6
0.35	7
0.40	8

1 Plot a graph of the data. Draw a line of best fit, and describe the correlation the graph shows.

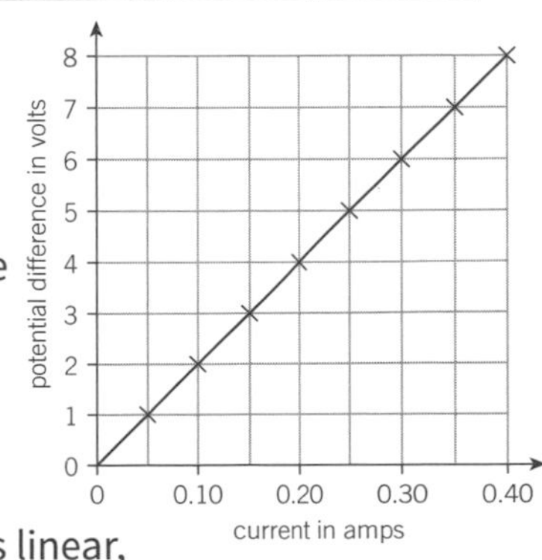

The graph is linear, and shows a perfect positive correlation.

2 Use the line of best fit to predict the potential difference if the current was 0.55 A.

= 11 V

Practice

1 The table below shows how the concentration of HCl changed over the course of a reaction.

Plot a graph of the data. Draw a line of best fit.

Time in s	Concentration of HCl in mol/dm^3
25	1.25
50	0.95
75	0.80
100	0.70
125	0.60
150	0.55
175	0.50
200	0.50

Practice

Exam-style questions

01 A student sets up a simple motor.

They connect the motor to a battery as shown in **Figure 1**.

Figure 1

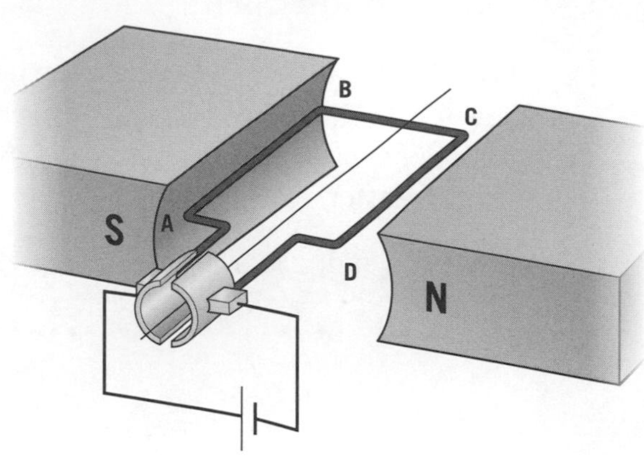

01.1 Describe the direction of the current in the coil of wire in terms of **A**, **B**, **C**, and **D**. **[1 mark]**

> **Exam Tip**
>
> You are asked to use the letters in your answer, so make sure you do!

01.2 Compare the force on side **AB** with the force on side **CD**.

Use your answer to describe which way the coil spins. **[3 marks]**

> **Exam Tip**
>
> You will need to use Fleming's Left Hand rule for **01.2**. The only way to find the answer is by getting that hand in shape.

01.3 Explain why there is no force on side **BC**. **[2 marks]**

01.4 When the coil is vertical, the contacts are no longer in contact with the battery and a current no longer flows in the coil.

Explain why the coil continues to move. **[2 marks]**

02 A student uses a magnetic field sensor to investigate how the magnetic field strength varies with distance from an electric wire.

02.1 When there is no current flowing in the wire the field sensor measures a magnetic field of 49 μT.

1 μT = 10^{-6} T

Write down what the sensor is measuring when there is no current passing through the wire.

Write down the name of this type of error. **[2 marks]**

Exam Tip

μT are micro Tesla's. There are 1 000 000 in one Tesla.

02.2 The student records their data in **Table 1**.

Table 1

Distance in cm	Magnetic field strength in mT			
	1	2	3	Mean
1	0.190	0.210	0.210	0.203
2	0.095	0.105	0.105	0.102
3	0.063	0.070	0.070	0.068
4	0.048	0.053	0.053	0.051

Suggest what the student should do about the sensor reading in **02.1** before recording the data. **[1 mark]**

Exam Tip

The error will have shown up in every result that the student recorded.

02.3 Suggest why the student does not start measuring at zero centimetres. **[1 mark]**

02.4 The student concludes that the magnetic field strength is inversely proportional to the distance from the wire.

Use the data in **Table 1** to explain why. **[3 marks]**

03 A tool set contains a screwdriver. The screwdriver attracts the screw so that the person using it is less likely to lose it.

03.1 The end of the screwdriver is magnetic. The screw is an induced magnet.

Explain the difference between a permanent magnet and an induced magnet. **[2 marks]**

03.2 There is a magnetic field around the end of the screwdriver.

Predict whether there is a magnetic field around the screw when it is attached to the screwdriver.

Justify your answer. **[2 marks]**

03.3 On **Figure 2** write N (north) and S (south) in the blank boxes to show the induced poles on the screw. **[1 mark]**

Figure 2

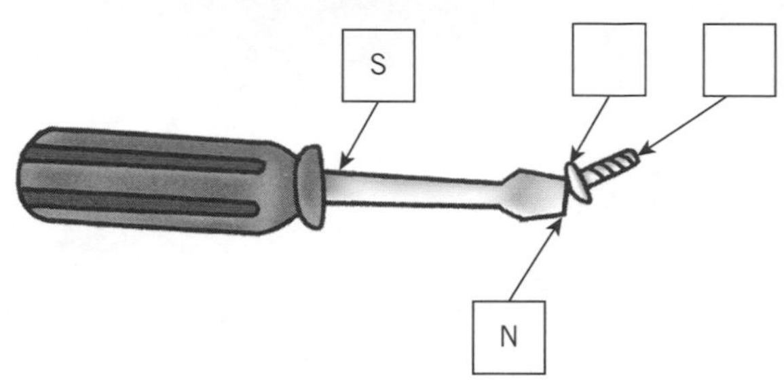

Remember, opposite poles attract.

03.4 The screw is put back into a box containing other screws.

Predict whether it will attract the other screws in the box.

Justify your answer. **[2 marks]**

04 A student investigates the strength of an electromagnet. They make a solenoid with different materials for the core and measure the mass of iron filings that the solenoid can pick up.

04.1 Write down the independent and dependent variables. **[2 marks]**

04.2 Write down **two** control variables. **[2 marks]**

04.3 The results are in **Table 2**.

Table 2

Material	Mass of iron filings in g			
	1	2	3	Mean
nickel alloy	0.10	1.0	0.10	0.10
steel	1.20	1.38	1.36	1.31
aluminium	0.03	0.02	0.02	0.02
iron	1.26	1.43	1.38	1.36

The control variables are the bit you need to keep the same to ensure it is a fair test.

One of the values is an outlier. Identify the outlier.

Write down what the student did about this outlier when they calculated the mean. **[2 marks]**

04.4 Nickel is a magnetic material.

Suggest whether the amount of nickel in the nickel alloy is large or small.

Give reasons for your answer. **[2 marks]**

04.5 The student learns that some high voltage machines, such as X-ray machines, use solenoids. These machines are operated by a switch that is not in the same circuit as the high voltage source (**Figure 3**).

Figure 3

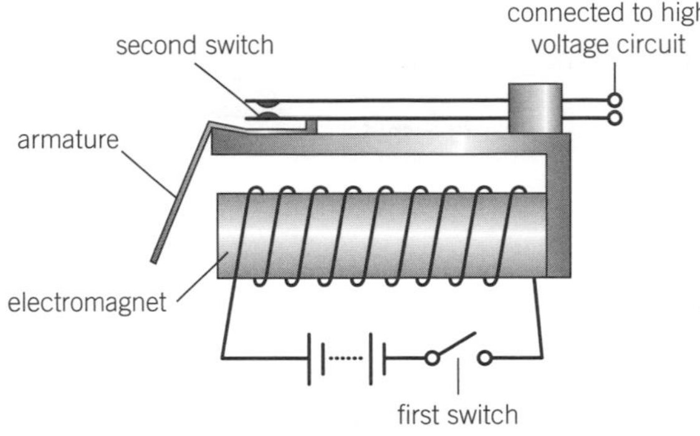

Suggest a suitable material for the core and armature. **[1 mark]**

04.6 Describe what happens when the first switch is pressed. **[3 marks]**

Exam Tip

Make sure you clearly indicate which switch and circuit you are talking about. If it is not clear you won't get the marks.

05 A student winds some wire around a wooden rod to make a coil and connects the coil to a battery.

05.1 Identify a hazard when doing this experiment. Suggest a method of reducing the risk of harm. **[2 marks]**

05.2 On solenoid **A** in **Figure 4**, draw lines to show the shape of the magnetic field around the coil. You do **not** need to draw arrows on the field lines. **[1 marks]**

Figure 4

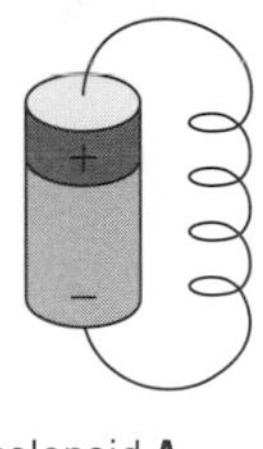

solenoid **A**

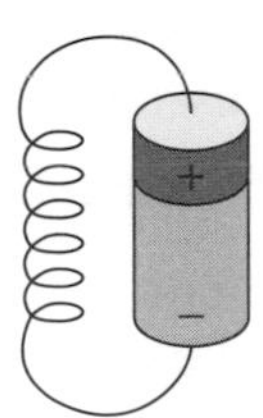

solenoid **B**

05.3 Write down which solenoid, **A** or **B**, has the strongest magnetic field around it. Give reasons for your answer. **[2 marks]**

05.4 The student takes a compass and places it in the centre of the solenoid. They move it up and down in the middle of the coil.

Will the compass needle move when it is in the solenoid? Justify your answer. **[2 marks]**

06 A student makes a simple motor. They use a piece of wire to make a coil and connect the coil to a battery. There is a current flowing in the coil but the coil does not spin.

06.1 Describe what the student needs to do to make the coil spin. Give reasons for your answer. **[2 marks]**

06.2 The student wonders if they can make a loudspeaker using the same equipment used for the motor.

Compare the construction of a simple direct current motor with the construction of a simple loudspeaker.

Suggest and explain what would happen if the student constructed a loudspeaker with the same equipment as the motor. **[6 marks]**

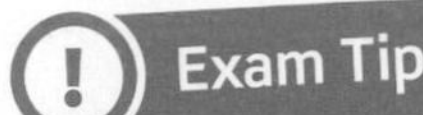

Moving coil loudspeakers and moving coil microphones operate in a very similar manner. It is worth learning this mechanism because it can be applied to a range of different answers.

07 There is a region around a magnet where there is a magnetic field.

07.1 Describe what is meant by magnetic field. **[1 mark]**

07.2 A student puts two magnets, **A** and **B**, together so that their north poles are facing each other.

Draw a diagram to show the magnetic field between the two magnets.

Draw arrows to show the direction of the magnetic field. **[2 marks]**

07.3 The magnetic field is stronger near the poles. Describe how this is shown on a diagram. **[1 mark]**

07.4 There is a 'neutral point' between the magnets where you can place a piece of magnetic material and it will not move.

Suggest where the point is in relation to the two magnets, **A** and **B**, when

- the magnets are equally strong
- magnet **B** is stronger than magnet **A**.

Explain your answers. **[5 marks]**

Exam Tip

A carefully annotated diagram can get marks in **07.4**. If you struggle to explain things with words then this is a perfect question to use a diagram.

08 There are craters produced from the impact of asteroids with the Earth's surface. Astronomers try to work out the position of asteroids that could collide with the Earth using telescopes.

They send a pulse of radio waves to an asteroid. The radio waves are reflected by the asteroid.

They use the time it takes to detect the reflected radio waves to work out the distance to the asteroid.

08.1 A telescope detects a reflected wave from an asteroid 0.2 s after a pulse of radio waves is emitted.

Calculate the distance to the asteroid. The speed of electromagnetic radiation is 3.0×10^8 m/s. **[4 marks]**

08.2 The uncertainty in the measurement of time is 1×10^{-4} s. Define uncertainty in this context. **[1 mark]**

08.3 The asteroid is moving. Suggest how the astronomers could use pulses of radio waves to calculate the speed of the asteroid. **[3 marks]**

Exam Tip

If you're not used to using standard form in science then take time to look it up in your maths books and practice it. It will come up in an exam somewhere.

09 A student connects a long thin strip of aluminium foil in a circuit with a battery. They lay the foil on the desk and bring a very strong magnet close to the foil. The foil moves.

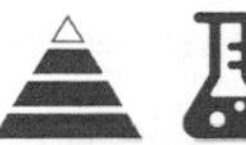

Exam Tip

Data loggers are a great way to increase the resolution of results. They can take results continuously and feed them to a computer. They can also draw the graph for you!

09.1 Suggest why the foil moves. **[1 mark]**

09.2 The student notices that as they move the magnet further away from the foil it no longer moves. Explain why. **[2 marks]**

09.3 The student also has a magnetic field sensor attached to a datalogger which measures magnetic field strength. The student wants to collect data to find the relationship between the magnetic field strength around the foil and the distance from the foil.

Describe a method of collecting data to find this relationship. **[3 marks]**

09.4 Sketch the graph the student would plot with the data they collected. Explain the shape of the graph. **[2 marks]**

Exam Tip

09.4 is a two mark question. One mark will be for the correct shape of the line on the graph and the other for explaining why you drew it like that.

10 A scientist notices that a compass needle is deflected when they turn on a circuit containing a battery and a wire.

10.1 Which direction was the compass pointing before the scientist turned on the circuit?

Explain why the compass points in this direction. **[2 marks]**

10.2 Compare the strength of the Earth's magnetic field with the strength of the magnetic field around the wire.

Give reasons with your answer. **[2 marks]**

10.3 Choose the correct description of a compass needle. Choose **one** answer. **[1 mark]**

a compass needle always points to the south pole

a compass needle is a magnet

a compass needle can be made from any metal

11 The Earth has a magnetic field.

11.1 Explain the difference between geographic North and magnetic north. **[2 marks]**

11.2 In 1600 William Gilbert published a paper in which he proposed that the Earth behaved like a giant magnet. Gilbert used a small physical model of the Earth with a magnet inside it.

Suggest the equipment he could have used to develop his ideas about the Earth. **[2 marks]**

11.3 Since Gilbert's idea was published, different scientists have proposed different models to explain the mechanism behind the bar magnet.

Suggest **one** reason why the models have changed. **[2 marks]**

11.4 Give **one** reason why scientists publish papers in journals. **[1 mark]**

12 A student uses a datalogger to measure the speed of rotation of a motor as she changes the current through it.

Her results are shown in **Table 3**.

Table 3

Current in A	Speed in rev/s
0.00	0
0.05	45
0.10	100
0.15	183
0.17	227

12.1 Plot the data in **Table 3** onto **Figure 5**. Draw a line of best fit.

[3 marks]

Figure 5

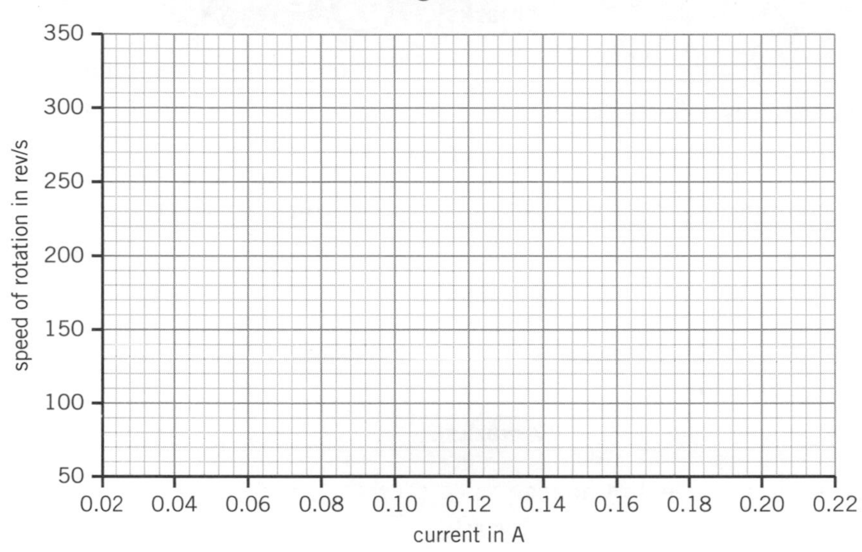

Exam Tip

Always uses crosses to plot points and draw a line of best fit.

12.2 Describe the relationship between the current and the speed.

[2 marks]

12.3 The student wonders why the speed changes. They write down the equation that relates force, magnetic field, current, and length. They also write down the equation that relates force, mass, and acceleration.

Use these two equations to explain the relationship that you described in **14.2**. **[5 marks]**

13 A teacher sets up a demonstration to show how an electric motor works. They suspend a light metal current-carrying rod in a magnetic field as shown in **Figure 6**.

Figure 6

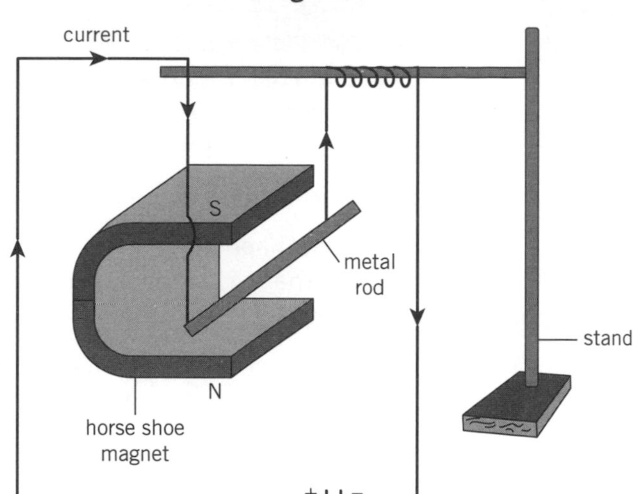

Exam Tip

This is the same set-up as in **Figure 1**, even if it does look very different. This is seeing if you can take what you know and apply it to a new situation. You will definitely have to do this in your exam. If you think a question in your exam is asking you something that isn't on the specification, it isn't! Look for the science behind the question – this will be familiar to you.

13.1 Write down the direction that the rod will move. **[1 mark]**

13.2 The current in the rod is 1.2 A. The length of the rod is 10 cm. The rod experiences a force of 2×10^{-3} N.

Calculate the strength of the magnetic field between the poles of the horseshoe magnet.

Use an equation from the *Physics Equations Sheet*.

Give the unit with your answer. **[4 marks]**

13.3 Magnetic field lines are like elastic bands. When they are stretched they try to return to their original shape. This produces a force.

Compare the shape of the magnetic fields between the poles of the horseshoe magnet and around the rod **Figure 6**.

Suggest what happens to the fields when they combine and why the rod moves. **[5 marks]**

14 A student watches a video that shows that some insects can detect ultraviolet light.

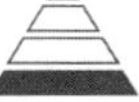

14.1 Explain why you cannot write down a single number for the wavelength of UV light. **[1 mark]**

14.2 The video explains that humans can see red light but some insects cannot.

Describe the difference between red light and ultraviolet light in terms of frequency and wavelength. **[2 marks]**

14.3 Ultraviolet light is hazardous to the human body. Describe **one** reason why. **[1 mark]**

14.4 Describe **one** use of ultraviolet light. **[1 mark]**

Exam Tips

You need to give two differences between red light and UV light. You can structure your answer like this:

The frequency of red light is ... whereas for ultraviolet it is ... .

The wavelength of red light is ... whereas for ultraviolet it is ... .

Knowledge

P18 Induced potential and transformers

The generator effect

A potential difference is **induced** (created) across the ends of a conductor if

- the conductor is moving relative to a magnetic field
- the magnetic field around the conductor changes.

In a complete circuit, there will be an induced current. The **generator effect** can be seen by:

1 moving a wire in a magnet field so that it cuts across the field lines

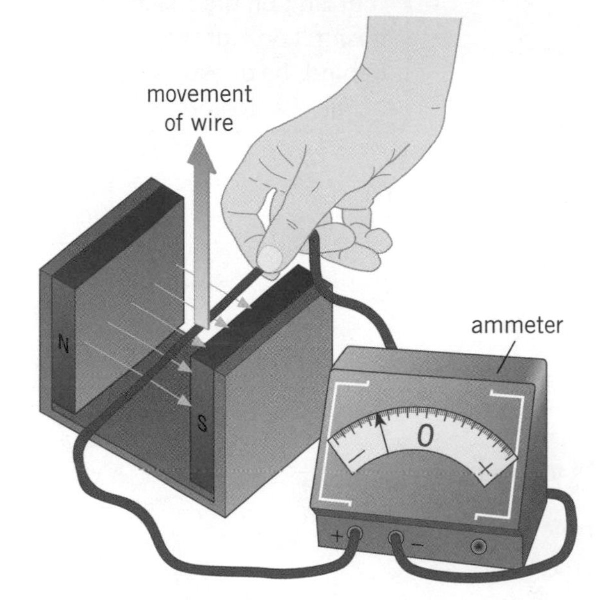

Factors affecting induced current

The direction of the induced potential difference and induced current reverses if

- the movement of the wire or magnet is reversed
- the **polarity** of the magnet is reversed.

If this reversal happens repeatedly, an alternating current/alternating potential difference is produced.

The induced potential difference/current will *increase* if the speed of movement, strength of the magnetic field or number of turns in the coil are increased.

An induced current will generate a magnetic field around the conductor that always opposes the original change producing it. This acts to slow down or stop any movement or change in magnetic field.

2 moving a magnet in and out of a coil of wire.

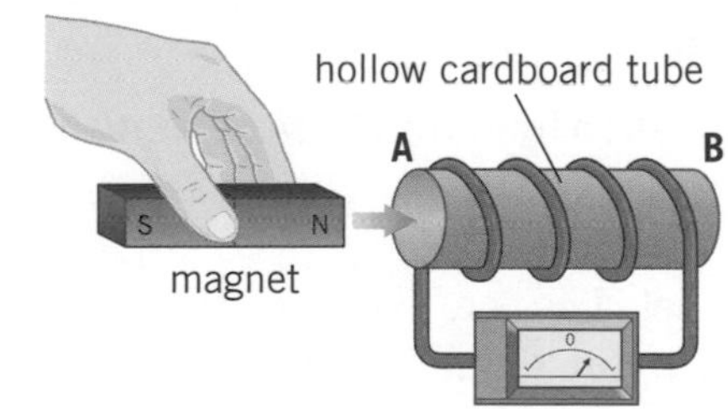

Transformers

A **transformer** is a device that can change the size of an alternating potential difference.

A basic transformer consists of two coils wound round an iron core.

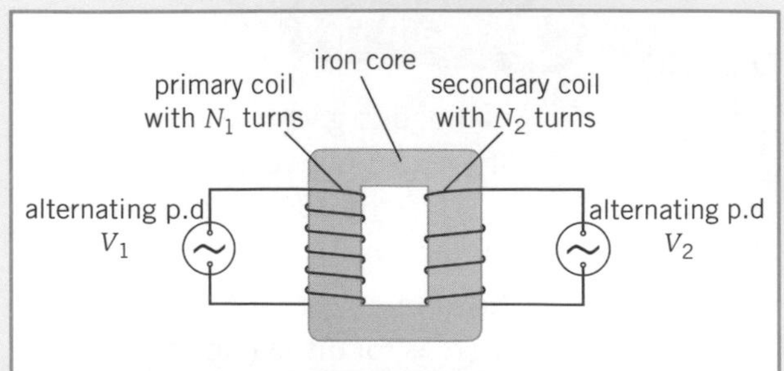

Iron is used for the core because it is easily magnetised and demagnetised.

How a transformer works

- An alternating current passes through the primary coil producing an alternating magnetic field in the iron core.
- The alternating magnetic field in the iron core induces an alternating p.d. in the secondary coil.

The ratio of potential differences across the primary and secondary coils is the same as the ratio of number of turns on each coil:

$$\frac{\text{potential difference across primary coil (V)}}{\text{potential difference across secondary coil (V)}} = \frac{\text{number of turns on primary coil}}{\text{number of turns on secondary coil}}$$

$$\frac{V_p}{V_s} = \frac{N_p}{N_s}$$

Uses of the generator effect

Alternators

When the coil is made to turn in the magnetic field, a current is induced in the coil.

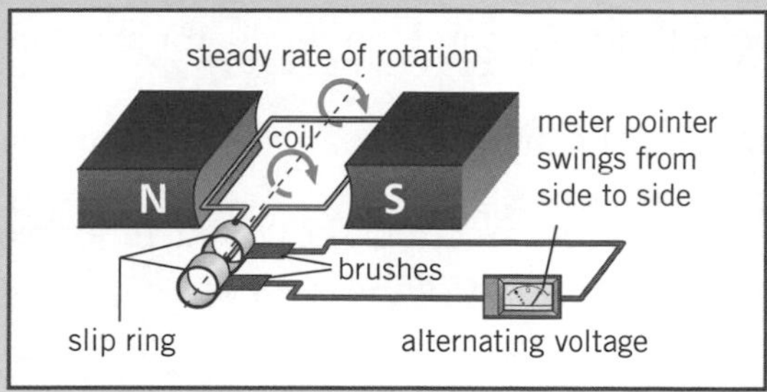

The ends of the coil are each connected to a slip ring. This keeps a continuous connection with the coil and lets the current flow out of the coil, through the brushes, and into a circuit.

The current changes direction in the coil every half turn:

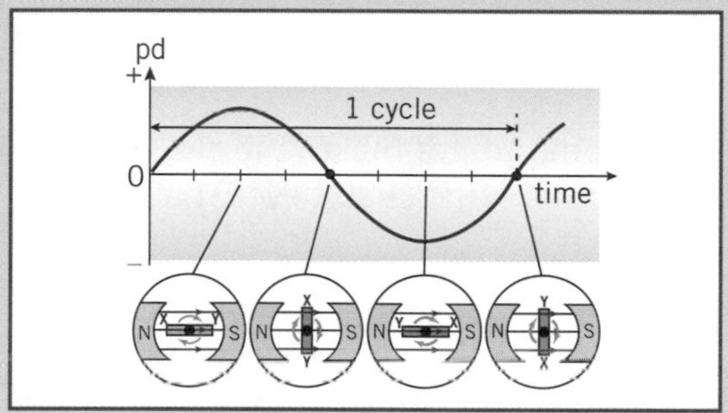

induced p.d. = maximum

when the plane of the coil lies parallel to the magnetic field lines

induced p.d. = 0

when the plane of the coil is perpendicular to the magnetic field lines

Dynamos

The **split-ring commutator** makes the ends of the coil swap contacts with the circuit every half turn, so the current always flows in the same direction relative to the magnetic field.

The induced p.d. in a **dynamo** varies from zero to a maximum value twice each cycle but never goes in the opposite direction.

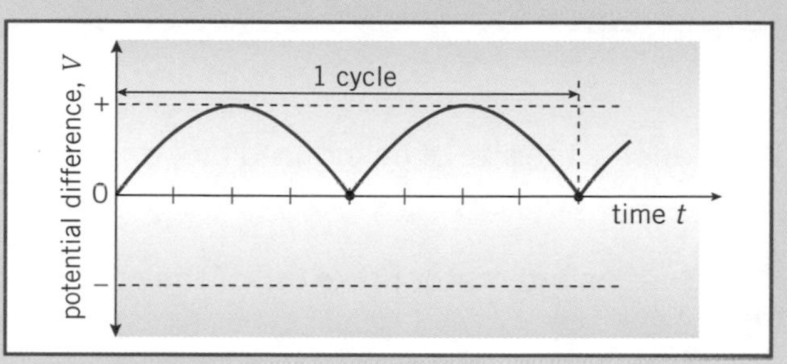

Microphones

A microphone uses the generator effect to convert variations in pressure in a sound wave into variations in electrical current.

- The coil is attached to a diaphragm.
- The coil is placed inside a permanent magnet.
- When a sound wave hits the diaphragm, it vibrates, making the coil vibrate inside the magnetic field of the magnet.
- This induces a current in the coil.
- The alternating p.d. induced in the coil has the same frequency as the sound waves which make the diaphragm vibrate.

If a transformer is 100% efficient: *power output of secondary coil = power input to primary coil* $V_s \times I_s = V_p \times I_p$

Remember that $P = V \times I$

In a step-up transformer:

- there are more turns on the secondary coil than on the primary coil so $V_s > V_p$
- current in the secondary coil I_s must be less than the current in the primary coil I_p for the power to be the same.

In a step-down transformer:

- there are fewer turns on the secondary coil than on the primary coil (N_p) so $V_s < V_p$
- current in the secondary coil I_s must be greater than current in the primary coil I_p for power to be the same.

Make sure you can write a definition for these key terms.

alternator | dynamo | generator effect | induced | polarity | split-ring commutator | transformer

Retrieval

Learn the answers to the questions below then cover the answers column with a piece of paper and write as many as you can. Check and repeat.

	P18 questions	Answers
1	Give two ways that a potential difference (p.d.) is induced in a conductor.	• conductor moving relative to a magnetic field • magnetic field around the conductor changing
2	What is another name for the generator effect?	electromagnetic induction
3	How can the direction of an induced p.d. or current be reversed?	• reverse the movement of the conductor or magnet • reverse the polarity of the magnet
4	What affects the size of an induced p.d. or current?	• speed of movement • strength of the magnetic field • number of turns in the coil
5	How does an alternator make use of the generator effect?	produces alternating current
6	Which kind of current is produced by the generator effect in a dynamo?	direct current
7	How does a dynamo keep the current flowing in the same direction?	a split-ring commutator makes the ends of the coil swap contacts with the circuit every half turn
8	When is the induced p.d. produced by an alternator at its maximum?	when the plane of the coil lies parallel to the magnetic field lines
9	How does a microphone use the generator effect?	converts pressure variations in sound waves into variations in electrical current
10	What does a transformer do?	changes the magnitude of the alternating p.d.
11	What does a basic transformer consist of?	two coils of wire wound around an iron core
12	Why is iron used for the core of a transformer?	iron is easily magnetised and demagnetised
13	How does a transformer work?	alternating p.d. across the primary coil produces an alternating magnetic field in the iron core which induces an alternating p.d. in the secondary coil
14	How is the ratio of p.d. across the primary coil and secondary coil related to the ratio of the number of turns on each coil?	the ratios are the same $\frac{V_p}{V_s} = \frac{N_p}{N_s}$
15	Which type of transformer has more turns on its secondary coil than on its primary coil?	step-up transformer
16	What is the efficiency of a transformer if the power output from the secondary coil is the same as the power input to the primary coil?	100%

Put paper here

Now go back and use the questions below to check your knowledge from previous chapters.

Previous questions / Answers

	Previous questions	Answers
1	How does a lever reduce the amount of force needed to create a particular sized moment?	by increasing the distance from the pivot
2	What will an object placed in a fluid do if its weight is greater than the upthrust?	sink
3	What happens to the drag on an object as its speed increases?	the drag increases
4	What does Newton's Third Law say?	when two objects interact they exert equal and opposite forces on each other
5	What does $m\,\Delta v$ stand for?	change in momentum
6	Which types of EM waves are harmful to the human body?	ultraviolet, X-rays, gamma rays
7	What is the frequency of a wave?	number of waves passing a fixed point per second
8	What are the properties of P-waves?	longitudinal, travel through liquids and solids
9	In which direction do magnetic field lines always point?	north to south

Put paper here

Put paper here

Put paper here

Maths Skills

Practise your maths skills using the worked example and practice questions below.

Graphs

Once you have plotted a graph, you often need to find the gradient of the graph. The gradient is another word for slope.

The gradient of a straight line graph can be calculated using:

$$\text{gradient} = \frac{\text{change in } y}{\text{change in } x}$$

When a graph is curved, you can find the gradient of the curve at a specific point. You do this by drawing a straight line tangent at that point and then calculating the gradient of the straight line.

Worked Example

Calculate the deceleration of the car at 300 s.

Step 1 Draw a straight-line tangent at the curve.

Step 2: Pick two points on the tangent and calculate the gradient of the tangent using:

$$\text{gradient} = \frac{\text{change in } y}{\text{change in } x} = \frac{1000 - 600}{150 - 300} = -2.6\ \text{m/s}$$

Practice

1 Use the distance–time graph to determine the speed of the car.

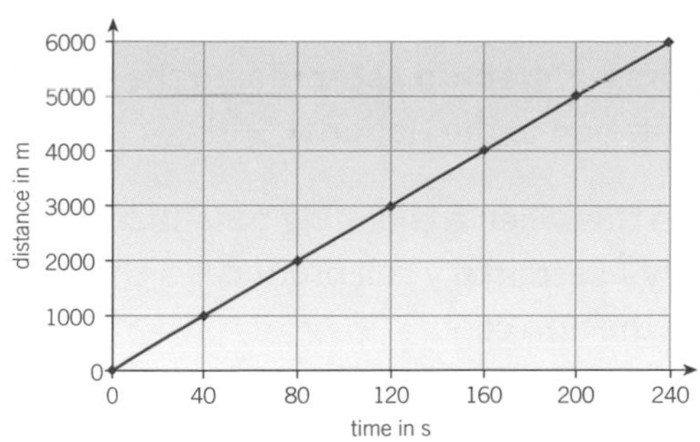

2 The distance–time graph shows a car as it approaches a sign.

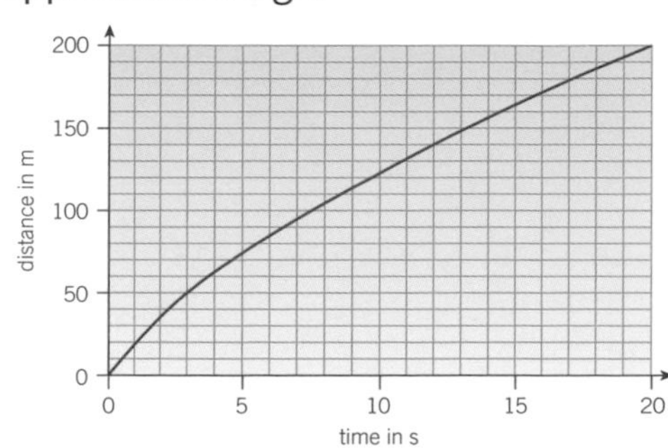

Determine the speed of the car at 10 seconds.

Practice

Exam-style questions

01 There are many devices that rely on a potential difference being induced.

01.1 Describe what is meant by an induced potential difference. **[1 mark]**

____

____

01.2 Draw **one** line from each statement to the correct equipment to complete the sentences. **[3 marks]**

Statement	Equipment
A device that produces an alternating potential difference is a…	…dynamo.
A device that changes the potential difference is a…	…microphone.
A device that produces a direct potential difference is a…	…generator.
A device that changes a sound wave to an electrical signal is a…	…transformer.

Exam Tip

Draw clear lines. If the examiner is confused about where your lines go they can't give you any marks.

01.3 The potential difference that is used by domestic appliances is produced by a generator in a power station. The power station is connected to the National Grid.

Circle the correct words or phrases in the sentences below. **[4 marks]**

The type of generator that is found in a power station is an **alternator** / **dynamo**.

Transformers **step-up** / **step-down** the potential difference from a power station to the National Grid.

They do this when a changing potential difference in the **primary** / **secondary** coil produces a changing **current** / **magnetic field** in the iron core.

Exam Tip

Approach this one sentence at a time. If you can't select between the set of words in the first sentence, don't give up. Try the next one.

02 A student has a magnet and a coil of wire that is connected to a voltmeter.

They observe that when they move the magnet into the coil it produces a potential difference that is positive.

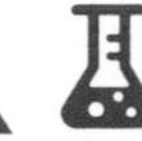

02.1 Describe how the student can increase the potential difference produced. **[1 mark]**

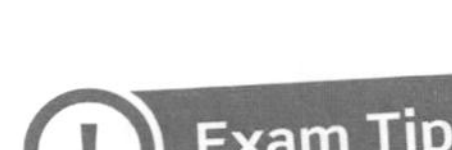

This is a one mark question so only needs a simple answer. Don't feel pressured to fill two lines with your answer.

02.2 Suggest how the student could model the production of alternating potential difference using the voltmeter, magnet, and coil.

Explain your answer. **[2 marks]**

02.3 When the student replaces the voltmeter with a resistor, a current flows in the coil when a magnet moves in or out of the coil.

The coil becomes an electromagnet.

Describe the link between the direction of the magnetic field inside the coil and the direction of the magnetic field of the moving magnet. **[1 mark]**

03 A student lies on a plank of wood that is supported at one end by a brick. The other end of the plank is on a scale.

03.1 Draw a free body diagram for the student. **[3 marks]**

03.2 Use Newton's First Law to explain why the student is stationary. **[2 marks]**

03.3 Using the information below, draw a diagram that shows the clockwise and anticlockwise moments acting on the plank. Assume that the brick is acting like a pivot, and ignore the weight of the plank. Label the distances from the pivot.

- The distance from the brick to the scale is 2.0 m.
- The centre of mass of the student is 1.2 m from the brick.
- The scale reads 400 N. **[4 marks]**

03.4 Use the law of moments to calculate the weight of the student. Give your answer to two significant figures. **[3 marks]**

> **! Exam Tip**
>
> Significant figures is a required maths skill across biology, chemistry, and physics. Make sure you are confident using them.

03.5 Write down the equation that links mass, weight, and gravitational field strength. **[1 mark]**

03.6 Calculate the mass of the student. Gravitational field strength = 9.81 N/kg. **[2 marks]**

04 A student is investigating the effect of changing the number of turns in the secondary coil of a transformer on the induced potential difference across the secondary coil.

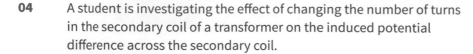

04.1 Describe what happens in a transformer to produce an induced potential difference. **[2 marks]**

04.2 The data from her experiment are shown in **Table 2**.

Table 2

Number of turns on the secondary coil	Induced p.d. in V			
	Repeat 1	Repeat 2	Repeat 3	Mean
0	0.0	0.0	0.0	0.0
10	2.5	2.4	2.6	2.5
20	4.8	5.2	5.0	5.0
30	7.3	7.4	7.8	7.5

Give **two** control variables in this experiment. **[2 marks]**

04.3 Describe the relationship between the number of turns on the secondary coil and the induced potential difference. Use the data to justify your conclusion. **[2 marks]**

04.4 The student wants to connect a bulb that requires a potential difference of 3 V at the secondary coil.

Suggest the number of turns on the secondary coil that the student needs in order to light the bulb. Explain your reasoning. **[2 marks]**

05 A teacher shows a mobile phone charger with the cover removed. A student says:

'I can see that the charger contains a transformer.'

05.1 Identify what the student saw inside the charger that prompted them to make that observation. **[1 mark]**

05.2 The teacher gives the following data about the charger:

$N_p = 2000$ $\qquad$ $N_s = 100$ $\qquad$ $V_s = 12\,V$

Use the correct equation from the *Physics Equations Sheet* to calculate V_p. **[3 marks]**

05.3 The student says:

'This is a step-up transformer because N_p *is bigger than* N_s.*'*

Do you agree? Give reasons for your answer. **[2 marks]**

05.4 Explain why the core of the transformer is made of iron. **[2 marks]**

06 A coil of wire spins in a magnetic field, as shown in **Figure 1.**

Figure 1

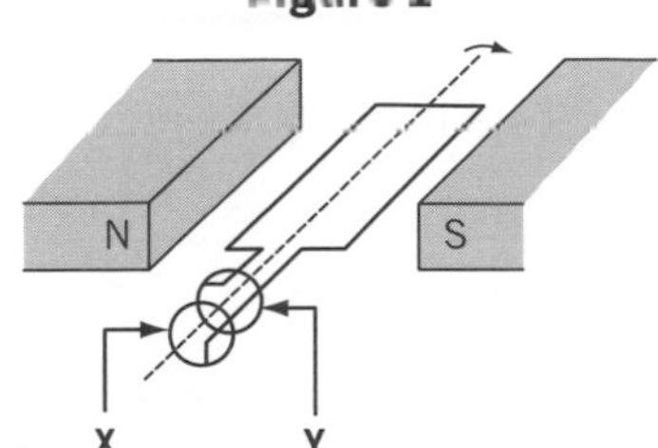

A student puts a voltmeter between points **X** and **Y**. They measure a changing potential difference when the coil spins.

06.1 Explain why a potential difference is produced. **[2 marks]**

06.2 A graph of potential difference against time is shown in **Figure 2**.

Figure 2

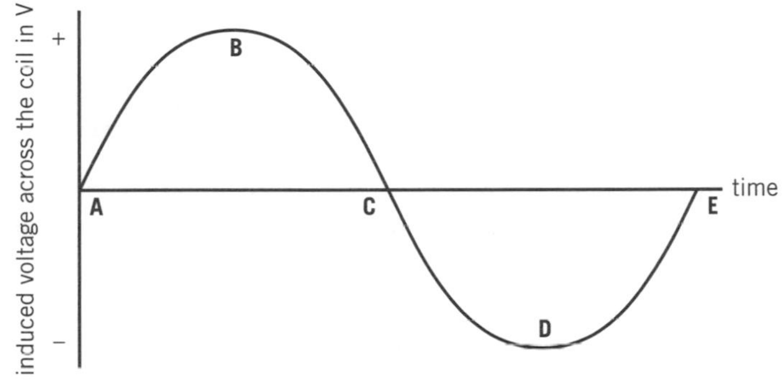

Exam Tip

Even though there are no numbers on this graph you still need to use data from it in your answer.

The coil in **Figure 1** is moving clockwise. Explain the shape of the graph. Suggest the **two** letters in **Figure 2** that correspond to position of the coil in **Figure 1**. Explain why there are two possible positions. **[4 marks]**

06.3 The student changes the permanent magnets for two electromagnets. Suggest changes that the student could make to the experimental set-up to produce the same graph as in **Figure 2** without spinning the coil. **[2 marks]**

07 A student looks at the a.c. adapter that they use for their laptop. The label on the adapter is shown in **Figure 7**.

Figure 7

a.c. adapter
Input: 100–240 V ∼ 1.8 A
Output: 19.5 V — 4.62 A

The adapter is connected to the mains and to the student's laptop.

07.1 Suggest what the following symbols mean in relation to the input and output current: — and ∼ **[2 marks]**

07.2 Write down the equation that links potential difference, current, and power in an electrical circuit. **[1 mark]**

07.3 Calculate the input power of the adapter. Use the value of the mains potential difference and the value of current given on the label.

Give your answer to two significant figures. Assume the adapter is being used in the UK. **[3 marks]**

07.4 Use the values given on the label to calculate the output power. **[2 marks]**

07.5 The student observes that the adapter gets hot when it is connected to the mains.

Compare the values of input and output power that you calculated in **07.3** and **07.4**.

Suggest how the student's observation can be explained. **[3 marks]**

> **Exam Tip**
> Think back to other topics on energy.

08 A student connects a dynamo to a datalogger. They spin the dynamo and look at the graph of potential difference against time that is produced by the datalogger.

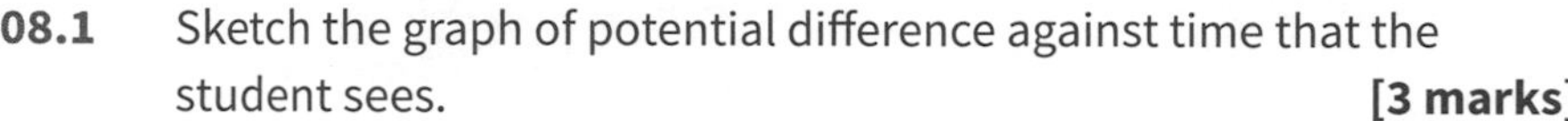

08.1 Sketch the graph of potential difference against time that the student sees. **[3 marks]**

> **Exam Tip**
> Sketching a graph means showing the general shape of the line and labelled axes. You don't have to plot points.

08.2 Explain how you can tell from the graph that the device is a dynamo. **[3 marks]**

08.3 Describe the changes to the graph that the student would see when they spin the coil in the opposite direction. **[1 mark]**

08.4 Describe the changes to the graph that the student would see when they spin the coil twice as fast. **[2 marks]**

09 The National Grid uses a variety of transformers. There are losses in energy due to heating.

09.1 Suggest why there is thermal energy loss when a transformer is in operation. **[2 marks]**

09.2 Write down the equation that links power, potential difference, and current in a circuit. **[1 mark]**

09.3 The output of a transformer supplies 120 MW of power to the National Grid at a potential difference of 400 kV.

Calculate the current in the secondary coil of a transformer connected to the National Grid. **[4 marks]**

> **Exam Tip**
> You can write the equation with words or symbols, whichever you find less confusing.

10 A teacher is demonstrating how a microphone works. They use a microphone to convert sound to electrical signals.

The signals are displayed using a device called an oscilloscope (**Figure 3**).

The teacher then shows a diagram of a moving-coil microphone (**Figure 4**).

Figure 3

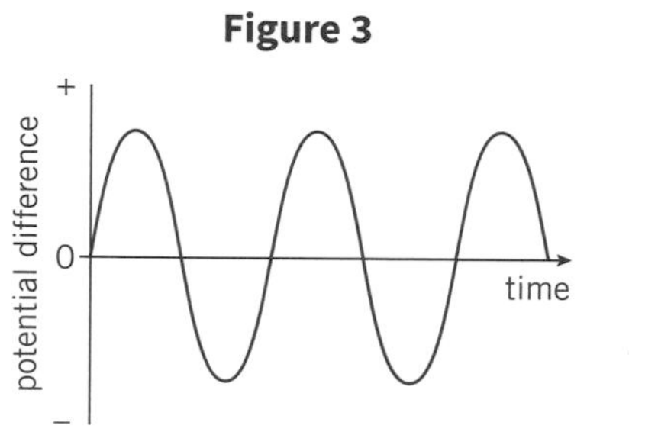

Figure 4

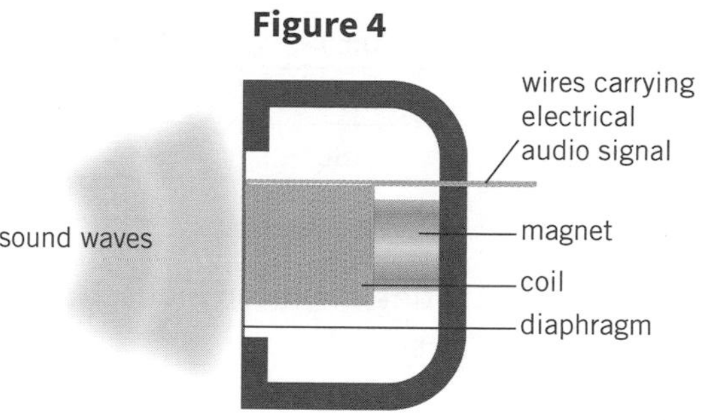

Describe how the pressure in a sound wave produces the output of the microphone on the screen. **[6 marks]**

11 A company making steel pipes uses ultrasound to find cracks in the pipes. Ultrasound has a frequency beyond the range of human hearing.

11.1 Write down the range of human hearing. **[1 mark]**

11.2 Write down the equation that links the speed, frequency, and wavelength of a wave. **[1 mark]**

11.3 Calculate the wavelength of ultrasound with a frequency of 400 kHz. The speed of sound in steel is 5000 m/s. **[3 marks]**

11.4 A technician uses ultrasound with a frequency of 400 kHz to measure the distance to the crack. The ultrasonic transducer detects a reflected pulse after 8.4×10^{-7} seconds.

Calculate the total distance travelled by the pulse. **[3 marks]**

11.5 Use your answer to **11.4** to calculate the distance to the crack in the pipe. **[2 marks]**

Exam Tip

The units for Hz are important to get correct. It is a capital H and a lower case z. Only Hz will get the marks. HZ will not get marks, hz will also not get marks.

12 A single piece of wire is moved between the poles of a magnet, as shown in **Figure 5**.

Figure 5

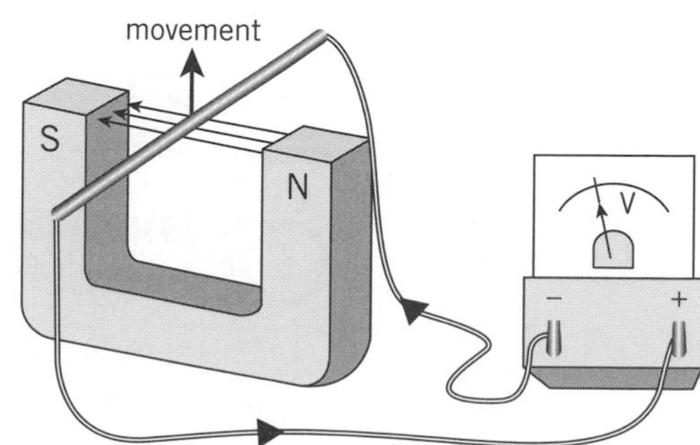

The speed of motion is increased and the induced potential difference is measured. The data are shown in **Table 4**.

Table 4

Speed in m/s	Induced p.d. in V			
	Repeat 1	Repeat 2	Repeat 3	Mean
0	0.00	0.00	0.00	0.00
1	0.70	0.67	0.68	0.68
2	0.90	1.01	1.20	1.04
3	1.60	1.80	1.45	1.62
4	1.85	2.00	1.90	

12.1 Complete **Table 4** by calculating the mean induced potential difference for a speed of 4 m/s. **[1 mark]**

Exam Tip

Give your answer to 2 decimal places. Always try to match the number of decimal places of your answer to the data given in the question.

12.2 Plot a graph of the data on **Figure 6**. Draw a line of best fit. **[2 marks]**

Figure 6

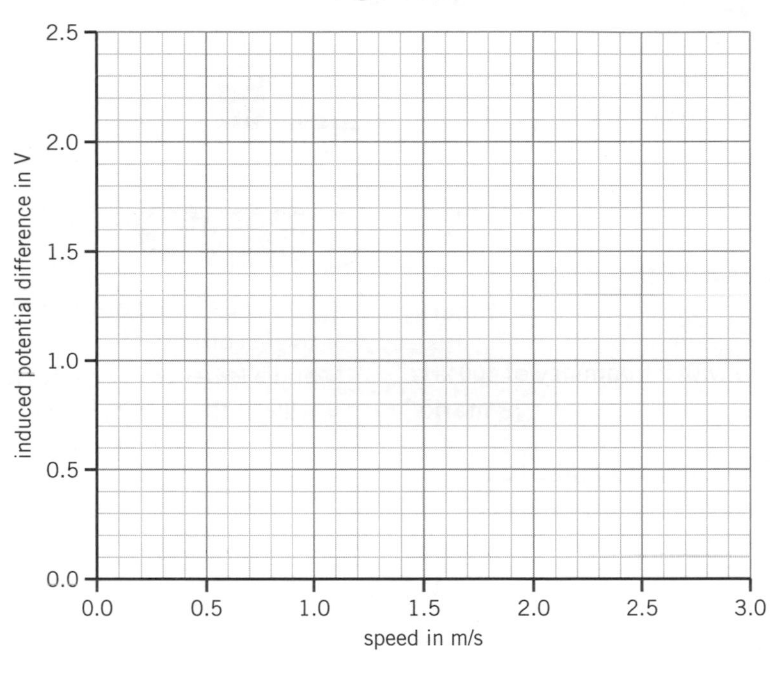

Exam Tip

Plot the mean induced potential difference, not each repeat.

12.3 Use **Figure 6** to find the potential difference that would be obtained at a speed of 1.5 m/s. **[1 mark]**

12.4 There is a spread in the data in **Table 4**. Name the type of error that produces this type of uncertainty in the data. **[1 mark]**

13 There are analogies between atmospheric pressure and water pressure. A student says:

'The way pressure changes when you go down a mountain is the same as the way pressure changes when you dive into the ocean.'

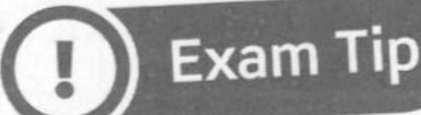

13.1 Describe a simple model of the Earth's atmosphere. **[2 marks]**

Exam Tip

This is only two marks so don't write a long and detailed answer.

13.2 Compare how the pressure changes when you come down a mountain with how the pressure changes when you dive down in an ocean. **[4 marks]**

13.3 Atmospheric pressure changes by 1 kPa for every change of 84 m. Water pressure changes by 1 kPa for every change of 0.1 m.

Use this data to calculate the ratio of the density of water to the density of air. **[2 marks]**

14 A motorboat is in the middle of a lake. It is stationary. When the engines are turned on the boat accelerates.

14.1 Draw a free body diagram for the boat while it is accelerating. Label the arrows in the diagram. **[4 marks]**

14.2 Write down the equation that links force, mass, and acceleration. **[1 mark]**

14.3 The engine provides a force of 3000 N. The mass of the boat is 860 kg.

Assume that there are no resistive forces acting on the boat. Calculate the acceleration of the boat. **[3 marks]**

14.4 The actual acceleration of the boat is 2.7 m/s^2.

Calculate the magnitude of the drag forces acting on the boat. **[5 marks]**

14.5 After the boat has finished accelerating it travels at a speed of 14 m/s.

Use an equation from the *Physics Equations Sheet* to calculate the distance travelled by the boat while it is accelerating.

Give your answer to two significant figures. **[3 marks]**

Exam Tip

There are lots of different parts to the calculations in question **14**. You might find it helpful to make a list of all the numbers you know before starting.

Knowledge

P19 Space

Our Solar System

Our **Solar System** is made up of the Sun (a star) and all the objects that orbit it, including:

- eight planets
- dwarf planets
- moons (natural **satellites**) that orbit planets
- asteroids
- comets

The Sun is located in the **Milky Way galaxy**, which contains billions of other stars.

Formation of stars

The Sun (and all other stars) was formed from a huge cloud of dust and gas (a **nebula**) pulled together by **gravitational attraction**.

Gravitational attraction between the particles of dust and gas causes them to merge together to form a **protostar**.

The protostar becomes denser as gravitational forces continue to pull it together, so the particles in the protostar collide more often.

More energy from the gravitational potential energy store of the particles is transferred to the thermal energy store, so the temperature of the protostar increases.

When the temperature is high enough, hydrogen nuclei fuse together to form helium nuclei.

These nuclear fusion reactions release huge amounts of energy and the protostar becomes a **main sequence star**.

This star is stable because the fusion reactions produce outwards forces which are in equilibrium with the gravitational forces pulling it inwards.

Life cycles of stars

All stars go through changes as part of a life cycle. The life cycle of a particular star is determined by its mass.

Starting as a nebula, stars with the same mass as the Sun, and more massive than the Sun, follow specific life cycles.

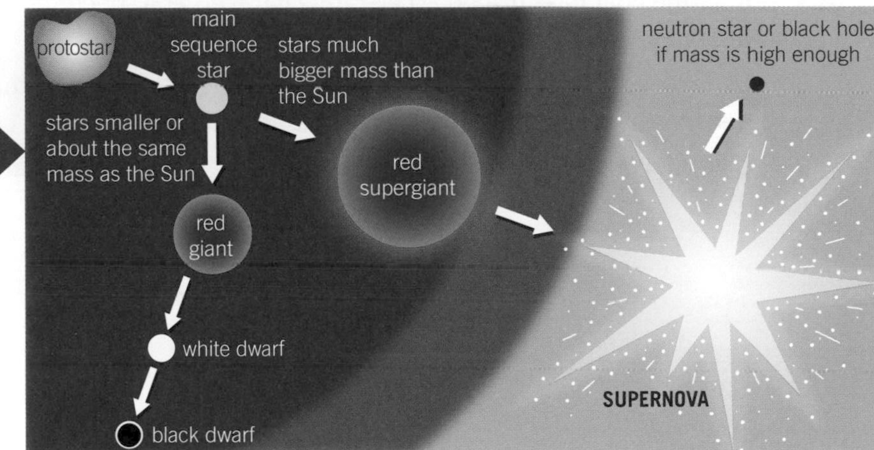

Formation of the elements

The nuclei for all the naturally occurring elements are produced by nuclear fusion in stars:

- hydrogen nuclei are fused together to form helium nuclei
- other small nuclei are formed in stars with large masses
- when a star becomes a red giant or red supergiant, helium, lithium, and other small nuclei are fused to form larger nuclei.

Elements heavier than iron require more energy to be produced, so are only produced when a massive star explodes (a **supernova**).

The elements produced in stars are distributed throughout the universe by massive stars going supernova.

Orbital motion and satellites

The Earth and other planets in the solar system **orbit** the Sun.

The Moon is a natural satellite that orbits the Earth, while other planets have other moons orbiting them.

The Earth also has artificial satellites orbiting it.

When one object orbits another, the less massive (smaller) object orbits the more massive (bigger) one.

Circular orbits

The Moon and the artificial satellites around the Earth move in circular orbits, while the orbits of the planets around the Sun are almost circular.

At any instant in time, the direction of the velocity of an object in a circular orbit is at right angles (perpendicular) to the direction of the resultant force acting on it.

Since the resultant force is at right angles to the velocity, it does not cause the object to speed up but only changes its direction.

An object moving in a circle is constantly changing direction, meaning it is constantly changing velocity (though not speed).

The object must therefore also be constantly accelerating, and so have a resultant force acting on it.

This resultant force is called the **centripetal force** and is always directed towards the centre of the circular orbit, so the acceleration of the object is always directed towards the centre.

For planets and satellites, gravity provides the resultant force that maintains their circular orbits.

Stable orbits

To stay in a stable orbit at a fixed distance from a larger object, the smaller object must move at a particular speed.

If the speed of an object in a stable orbit changes, the radius of the orbit must also change.

The slower the speed of an orbiting object, the bigger the radius of the circle it moves in.

Red-shift

Red-shift is the name given to the effect that makes the wavelengths of light *longer* if the light source is moving away from the observer.

Scientists have observed that the wavelengths of light from most distant galaxies are longer than expected – they are red-shifted.

This suggests that these galaxies are moving away from the Earth.

The further away galaxies are, the more their light is red-shifted, suggesting distant galaxies are moving away from Earth faster than close galaxies.

These observations suggest that the universe (space itself) is expanding.

Since 1998, scientists have observed light from supernovae that suggests distant galaxies are moving away faster and faster.

This indicates that the speed at which the universe is expanding is increasing.

Big Bang theory

Scientists used these observations to propose the **Big Bang theory** for the start of the universe.

The Big Bang theory suggests that the universe started off as an extremely small, hot, and dense object that exploded.

As well as the red-shift of light from galaxies, there is other evidence to support the Big Bang theory, like the existence of electromagnetic radiation that was produced just after the Big Bang.

Scientists still do not know or understand much about the universe or how it began.

For example, they think **dark energy** could be responsible for the acceleration of the expansion of the universe, and **dark matter** might provide the gravitational force holding galaxies together.

But these things are not understood, and models like the Big Bang theory may change following new observations.

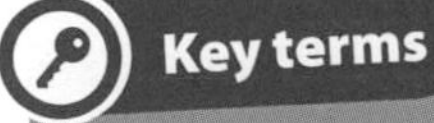

Make sure you can write a definition for these key terms.

Big Bang theory, centripetal force, dark energy, dark matter, gravitational attraction, main sequence star, Milky Way galaxy, nebula, orbit, protostar, red-shift, satellite, solar system, supernova

Learn the answers to the questions below then cover the answers column with a piece of paper and write as many as you can. Check and repeat.

	P19 questions	Answers
1	What are the main objects in our Solar System?	Sun, (eight) planets, dwarf planets, moons, asteroids, comets
2	What kind of object is the Sun?	star
3	Which galaxy is the Solar System in?	the Milky Way
4	What do all stars start off as?	huge cloud of gas and dust called a nebula
5	Which force is responsible for forming a protostar from a nebula?	gravity
6	What kind of reaction causes the expansion of a star?	nuclear fusion
7	How does a main sequence star remain stable?	fusion reactions produce outwards forces which balance the gravitational forces pulling it inwards
8	What determines the life cycle of a star?	mass
9	What is the life cycle of a star with about the same mass as the Sun?	protostar → main sequence star → red giant → white dwarf → black dwarf
10	What is the life cycle of a star with much more mass than the Sun?	protostar → main sequence star → red supergiant → supernova → neutron star or black hole (if mass big enough)
11	How are naturally occurring elements formed?	from nuclear fusion during the life cycle of stars
12	Which elements are only produced in a supernova?	elements heavier than iron
13	How are the elements distributed throughout the universe?	massive stars going supernova (exploding)
14	How does the force of gravity make objects in orbit change their velocity but not their speed?	gravity provides a centripetal force which keeps orbiting objects moving in a circle – they are constantly changing direction
15	To change the speed of an object in stable orbit, what factor must change?	radius of the orbit
16	What is red-shift?	wavelengths of light get longer if the light source is moving away from the observer
17	What evidence suggests that the universe is expanding?	light from more distant galaxies is more red-shifted, so more distant galaxies are moving away faster
18	What is the name of the scientific theory for the origin of the universe that suggests it started off as an extremely small, hot, and dense region?	the Big Bang theory

Put paper here

Now go back and use the questions below to check your knowledge from previous chapters.

Previous questions | Answers

	Previous questions	Answers
1	Are electromagnetic (EM) waves longitudinal or transverse waves?	transverse
2	What is thinking distance?	the distance vehicle travels during driver's.
3	What is another name for the generator effect?	electromagnetic induction
4	What is the motor effect?	when a conductor placed in a magnetic field experiences a force
5	What is the name of the scientific theory for the origin of the universe that suggests it started off as an extremely small, hot, and dense region?	the Big Bang theory
6	What causes the motor effect?	interaction between the magnetic field created by current in a wire and the magnetic field in which the wire is placed

Put paper here

Put paper here

Maths Skills

Practise your maths skills using the worked example and practice questions below.

Standard form	Worked Example	Practice
Standard form is a convenient way of writing very large or very small numbers. Numbers written in standard form must take the form: $A\times10^{n}$ 'A' is a decimal number between 1 and 10 (but not including 10). 'n' is a whole number and can be positive or negative. If n is positive, the number is greater than one. If n is negative, the number is less than one. For large numbers, the n is positive, and the decimal point is shifted to the left: $48\,000\,000 = 4.8\times10^{7}$ For small numbers, the n is negative, and the decimal point if shifted to the right: $0.000\,000\,48 = 4.8\times10^{-7}$ To convert numbers *from* standard form, move the decimal point back to the right for a positive n, and back to the left for a negative n.	The diameter of an atom is 0.000 000 000 1 m. Express this in standard form. **Answer:** as this is a small number, move the decimal point 10 places to the right to give a number between 1 and 10. $= 1.0\times10^{-10}$ The frequency of some radio waves is 27 500 000 00 Hz. Express this in standard form. **Answer:** this is a big number, so move the decimal point 9 places to the left. $= 2.75\times10^{9}$	**1** The wavelength of red light is around 7.0×10^{-7} m. Write this as a decimal number. **2** The distance from the Earth to the Sun is 15 000 000 000 0 m. Write this in standard form. **3** The estimated age of the universe is 4.32×10^{17} seconds. Write this as a decimal number.

Practice

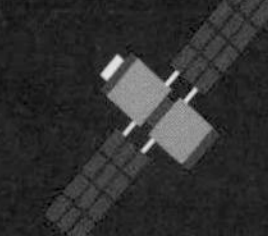

Exam-style questions

01 In August 2006 the International Astronomical Society changed the status of Pluto from planet to dwarf planet.

01.1 Complete the following statements using the words below. You may need to use the words once, more than once, or not at all. **[6 marks]**

Milky Way	moons	Sun	satellite
gravity	planets	Andromeda	centripetal

Both planets and dwarf planets orbit the ______________________.

Natural satellites, which are called ______________________, and artificial satellites orbit planets.

All ______________________ and ______________________ are in orbit because of the force of ______________________.

This force is an example of a ______________________ force.

Exam Tip

Read the instructions carefully: "You may need to use the words once, more than once, or not at all.", so don't panic if there is a word you haven't used.

01.2 Give **one** similarity and **one** difference between a planet and a moon. **[2 marks]**

Similarity: ____

____

Difference: ____

____

Exam Tip

There are lots of different types of planets, so be careful with your answer. Your answer must refer to all planets.

01.3 Give **two** similarities between an artificial and a natural satellite. **[2 marks]**

1 ____

____

2 ____

____

01.4 Give the name of the galaxy that the Sun and Earth are part of. **[1 mark]**

____

02 An astronomer makes an observation of the spectrum of light from a distant galaxy.

The galaxy is moving away from the Earth.

She compares the lines in the spectrum with the lines due to the same element produced in the Sun.

The line spectra are shown in **Figure 1**.

Figure 1

Sun

A B

distant galaxy

02.1 Write down which end of the spectrum, **A** or **B**, has the longer wavelength.

Explain your reasoning. **[2 marks]**

Exam Tip

You can still use data from the figure, even though there are no numbers given.

02.2 Describe the link between the position of the lines in the spectra and the speed of the galaxy. **[1 mark]**

02.3 Describe what the astronomer would observe if the galaxy was moving away at a smaller speed. **[1 mark]**

02.4 The astronomer observes that the star is inside our galaxy. The star is moving towards our solar system.

Describe what the astronomer would observe in terms of the spectrum. **[1 mark]**

Exam Tip

Stars that are moving towards us are blue-shifted. This is similar to red shift but in the other direction.

03 A teacher is explaining how astronomers work out the speed at which galaxies are moving away from our galaxy. The students watch a video showing how the pitch of a sound changes as the source of sound moves away. The sound gets lower in pitch. The teacher explains how this is analogous to the light being red-shifted.

03.1 Explain the link between the change in pitch and the change in colour. **[2 marks]**

03.2 Scientists have worked out that most galaxies are red-shifted, but stars within our galaxy can be red or blue-shifted. Suggest why there is a difference between the origin of the red shift when observing galaxies and the origin of the red shift when observing stars. **[2 marks]**

03.3 Suggest why astronomers would expect that the speed of expansion of the universe should be decreasing. **[1 mark]**

03.4 Describe how astronomers are explaining the observation that the speed of expansion of the universe is increasing. Suggest why they are not confident about this explanation. **[2 marks]**

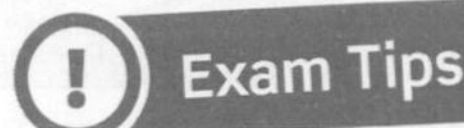

Galaxies have billions of stars within them.

For example, our Milky Way has 250 billion ± 150 billion stars in it, and most of them are moving away from the Earth.

04 A bar in a ripple tank makes ripples on the surface of the water.

04.1 Describe the motion of the bar that produces the ripples shown in **Figure 2**. Explain why the wave is transverse. **[2 marks]**

Figure 2

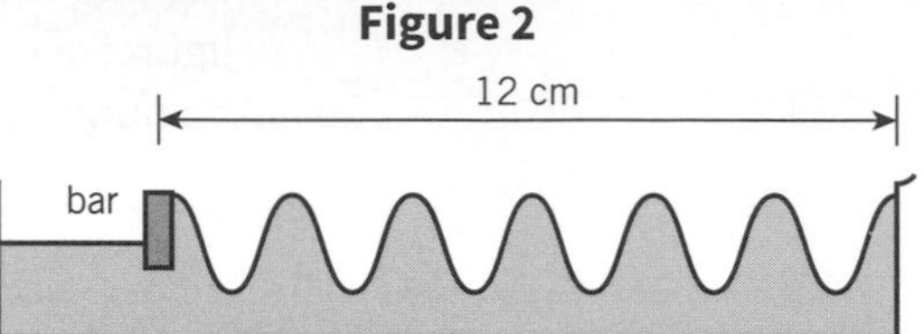

04.2 Write down the number of waves in the 12 cm shown in **Figure 2**. Use your answer to calculate the wavelength of the waves. **[2 marks]**

04.3 Write down the other quantity that a student needs to measure in order to use the wave equation to calculate the speed of the waves. **[1 mark]**

04.4 Suggest how you could make a precise measurement of this quantity. **[2 marks]**

Exam Tip

Instead of measuring the wavelength of one wave, it is better to measure the waves in a set distance and divide it by the number of waves.

This reduces errors from any variability in the waves.

05 Every star, including the Sun, has a life cycle. Here is the life cycle of Betelgeuse, a red supergiant.

A → protostar → main sequence → red supergiant → **B**

05.1 Circle the correct word or phrase in bold in the sentences below. **[3 marks]**

Betelgeuse is **about the same mass as / bigger than** the Sun

The word for **A** should be **neutron star / nebula**

The word for **B** should be **star / supernova**.

Exam Tip

The size of a star determines which life cycle it will follow – small and giant stars have different life cycles.

05.2 Give the name of the process that produces the energy to make a star shine. **[1 mark]**

05.3 The human body is made up of many different elements. Some people say 'Everyone is made of stardust'. Suggest how the elements that were formed in supernovae ended up on Earth. **[2 marks]**

06 Astronomers have discovered thousands of exoplanets. An exoplanet orbits a star that is not the Sun. One system of exoplanets is the Upsilon Andromedae system.

06.1 The planet Saffar has an orbit that is a circle. Describe the direction of the force on Saffar in relation to its motion. **[1 mark]**

06.2 The data for the solar system is shown in **Table 1**.

Table 1

Name of planet	Distance from star to planet in km	Time for one orbit in days
Saffar	8.91	4.62
Samh	124.35	241.00
Marjiriti	379.50	1276.00
Planet E	787.50	3848.00

Plot a graph of the time for one orbit against distance on **Figure 3**. **[2 marks]**

Figure 3

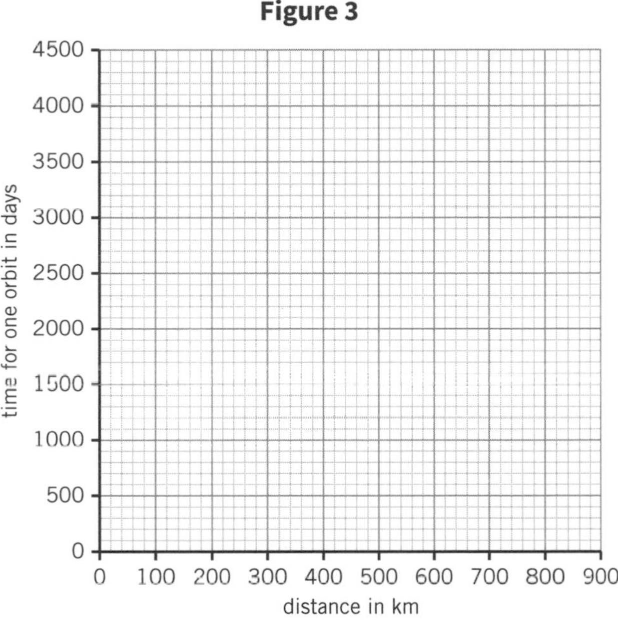

06.3 Astronomers discover a new exoplanet that is at a distance of 200 km from the star. Use **Figure 3** to estimate the expected time for one orbit of this exoplanet. Give the unit with your answer. **[2 marks]**

06.4 All the planets are in stable orbits. The radius of the orbit of Planet **E** is approximately twice that of Marjiriti. Use ratios, and the equation that links speed, distance, and time, to show that the speed of Planet **E** is about $\frac{2}{3}$ that of Marjiriti. **[3 marks]**

07 Scientists use the Big Bang theory to explain why the universe is expanding.

07.1 Describe what the Big Bang theory says about the beginning of the universe. **[3 marks]**

Exam Tips

Don't worry if you're never heard of these stars before.

This is testing if you can apply what you've learnt in class to a new context.

Always plot graph points as crosses, and include a line of best fit.

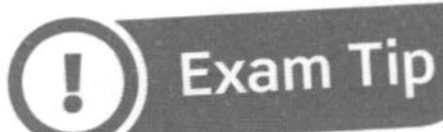

Draw construction lines on your graph to show your working.

07.2 Hubble found that galaxies that were further away had a larger recessional speed (the speed at which they were moving away). Explain how this provided evidence for the Big Bang theory. **[1 mark]**

07.3 Hubble also predicted that the recessional speed was proportional to their distance. Astronomers have now found that very distant galaxies are moving faster than expected. Sketch a graph to show how the speed of galaxies varies with distance if the speed of recession is proportional to the distance (label the line **P**). The gradient of the graph of velocity against distance is known as the Hubble constant. Explain why the units of the Hubble constant are 1/time. Add a line to your graph to show what happens if the speed of recession is increasing with distance (label the line **I**). **[6 marks]**

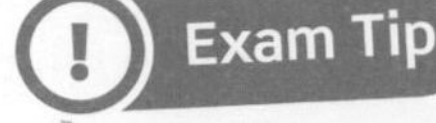

Sketched graphs need to show the general trend of the line and have labelled axis, but you do not need to plot individual points.

08 Our Sun is in a phase of its life cycle that will last for another 5 billion years.

08.1 Explain how the Sun formed and why it is now in a state of equilibrium. **[6 marks]**

08.2 A black hole is an object that is so dense that to escape from it requires a velocity that is greater than the velocity of light. Our Sun will not become a black hole. Compare the end of the life cycle of a star with a similar mass to the Sun with that of a larger mass star. **[6 marks]**

There is no limit on the number of six mark questions in an exam.

It may seem like a lot to have two next to each other, but this is a possibility!

09 The Earth is made of a mixture of elements. The atmosphere is 78% nitrogen and the crust is 62% oxygen. Until the 1920s, astronomers thought that the chemical composition of stars was the same as that of the Earth. In 1925, an astronomer called Cecilia Payne concluded from her research into the atmosphere of stars that stars were mainly composed of hydrogen with some helium.

09.1 Her conclusion was not accepted at the time. Suggest a reason why. **[1 mark]**

09.2 Some years later her conclusion was accepted. Suggest a reason for this change. **[1 mark]**

09.3 The process of fusion happens in stars. Complete the equation for the fusion of hydrogen. **[2 marks]**

$$ {}^{3}_{1}\mathrm{H} + {}^{2}_{1}\mathrm{H} \rightarrow {}^{\square}_{\square}\mathrm{He} + {}^{1}_{0}n $$

Exam Tip

This is a slight variation on decay equations you've seen, but treat it the same as the familiar equations.

09.4 The following elements are also produced by the process of fusion in stars:

$^{8}_{4}\mathrm{Be}$ $\quad$ $^{12}_{6}\mathrm{C}$

Use the reaction in **09.3** to suggest how $^{8}_{4}\mathrm{Be}$ and $^{12}_{6}\mathrm{C}$ are formed in stars. **[2 marks]**

09.5 It becomes increasingly difficult to fuse elements as they get more massive. Suggest why this is. **[3 marks]**

09.6 Name the element with the largest mass that is produced in the fusion process of stars. **[1 mark]**

10 A student models a satellite in an orbit by swinging a rubber bung in a horizontal circle, as shown in **Figure 4**. A piece of string connects the bung to some small slotted masses.

Figure 4

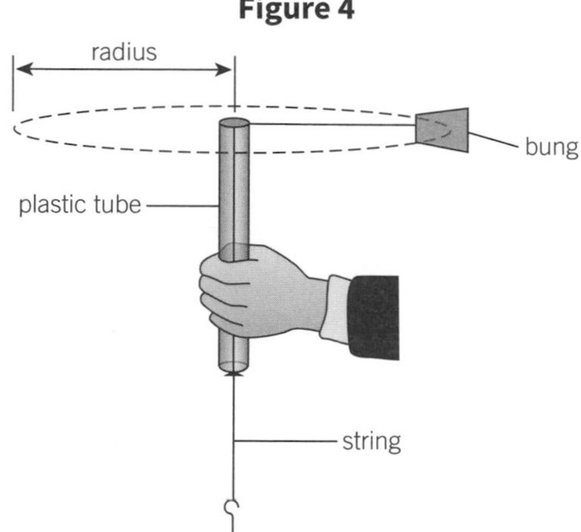

10.1 Write down what is analogous to the satellite. **[1 mark]**

> **Exam Tip**
> This is asking which part of the model in **Figure 4** is acting as the satellite.

10.2 Write down what is analogous to the force of gravity acting on the satellite. **[1 mark]**

10.3 The student measures the speed of the bung at different lengths of string for the same number of slotted masses. Describe **two** sources of uncertainty in the measurements. **[2 marks]**

> **Exam Tip**
> Think of all the different things that need to be controlled and how difficult it would be to control them.

10.4 Describe the relationship between the radius of the circular path and the speed of the bung. **[1 mark]**

10.5 Suggest another force that may act on a satellite to alter the speed of its orbit. **[1 mark]**

11 Astronomers measure the red shift of two galaxies. Galaxy **A** is further away than galaxy **B**.

11.1 Write down the letters of the true statements. **[2 marks]**

W: The red shift of galaxies is evidence for the Big Bang theory.

X: Galaxy **A** has a bigger red shift than galaxy **B**.

Y: Galaxy **B** is moving faster than galaxy **A**.

Z: The space between the galaxies is expanding.

> **Exam Tip**
> The number of marks can give you a clue to the number of true statements.

11.2 A student is modelling red shift. She draws a wave on a piece of elastic. Suggest how she can use the elastic to show how light is red-shifted. Explain what the model shows. **[2 marks]**

11.3 The model described in **11.2** is an example of a physical model. Give **two** reasons why scientists use models. **[2 marks]**

12 A student uses an application on her phone to measure displacement while she is using a lift. The phone detects the upwards direction as positive value. The graph produced by the phone is shown in **Figure 5.**

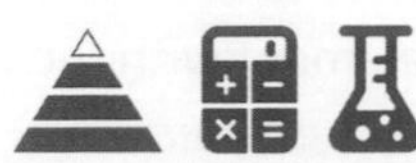

Exam Tip

There are three main sections to this graph. Divide it up and label each section with what is happening – this will help you in later questions.

Figure 5

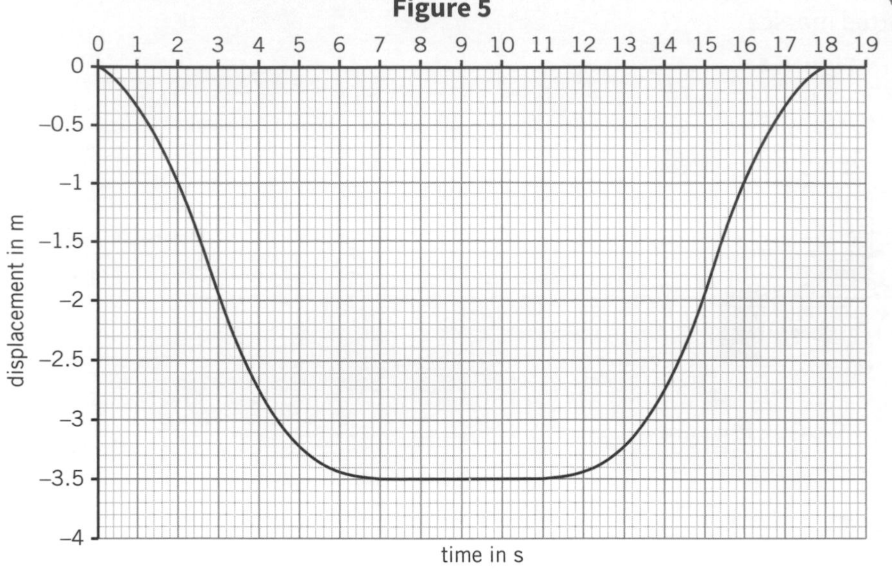

12.1 Describe the motion of the lift between 0 s and 4 s. **[3 marks]**

12.2 Describe the motion of the lift between 7 s and 10 s. **[1 mark]**

12.3 Calculate the velocity of the lift at 5 s. Show how you worked out your answer. **[3 marks]**

12.4 Sketch a velocity–time graph for the lift. **[4 marks]**

Exam Tip

Draw lines on your graph to help with the working out, and write down which equation you use.

13 A student blows up a balloon and seals it. He stands in the shallow end of a swimming pool and pushes the balloon into the water.

13.1 The balloon accelerates upwards when he lets it go. Explain why. **[2 marks]**

Exam Tip

Think about the forces involved.

13.2 The deep end of a swimming pool is 3 m deep. Calculate the pressure due to a column of water that is 3 m deep. Use an equation from the *Physics Equations Sheet*, and give an appropriate unit for your answer. The density of water is 1000 kg/m^3. Gravitational field strength = 9.8 N/kg. **[3 marks]**

13.3 A diver takes a pressure gauge to a depth of 3 m and finds that the reading is different to the pressure expected. Suggest how the measured reading will differ to the expected value, and give a reason for your answer. **[2 marks]**

14 A student is investigating what happens when a moving trolley collides with a stationary trolley and they stick together. They change the mass of the moving trolley and measure the velocity of the combined trolleys after the collision.

Exam Tip

The independent variable is the one that is changed, and the dependent variable is the one that is measured.

14.1 Identify the independent and dependent variables in this investigation. **[2 marks]**

14.2 Identify **two** control variables. **[2 marks]**

14.3 The student collects the data shown in **Table 2**.

Table 2

Mass of moving trolley in g	Combined final velocity in m/s			
	Repeat 1	Repeat 2	Repeat 3	Mean
100	0.50	0.80	0.60	0.63
200	0.90	1.10	1.20	1.07
300	1.20	1.30	1.30	1.27
400	1.34	1.35	1.34	1.34
500	1.41	1.43	1.43	1.42

Use **Table 2** to determine the uncertainty in the measurements of combined final velocity. Explain your method. **[2 marks]**

14.4 Plot a graph of the data on the axes in **Figure 6**. Suggest whether or not the line of best fit should go through (0, 0). Explain your answer. **[4 marks]**

Figure 6

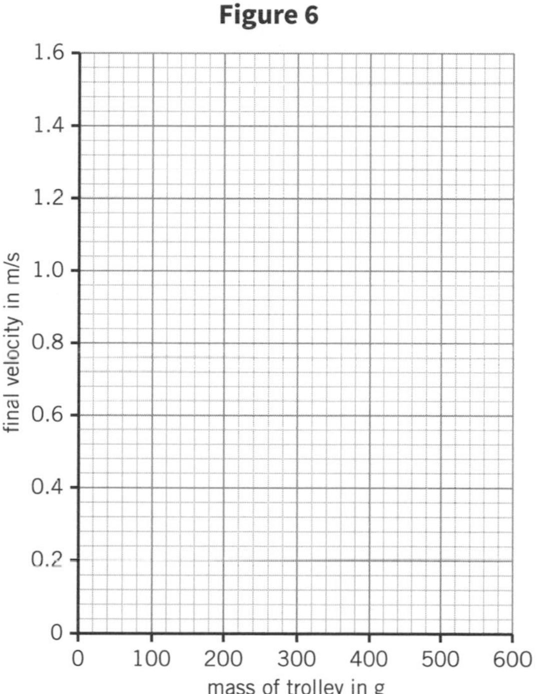

Exam Tip

Your line must fit the data points that are drawn – don't draw the line you expect to see.

14.5 The mass of the stationary trolley is 200 g. The initial velocity of the moving trolley is 2.0 m/s. Use the principle of conservation of momentum to calculate the predicted final velocity for a trolley with a mass of 100 g. Calculate the percentage error between the predicted and the measured value. Give your answer to an appropriate number of significant figures. **[6 marks]**

Physics equations

You need to be able to recall the following equations in your exams.

	Questions	Answers	
1	What is the equation for weight?	weight = mass × gravitational field strength (g)	$W = m\,g$
2	What is the equation for work done?	work done = force × distance (along the line of action of the force)	$W = F\,s$
3	What is the equation for force on a spring?	force applied to a spring = spring constant × extension	$F = k\,e$
4	What is the equation for the moment of a force?	moment of a force = force × distance (normal to direction of force)	$M = F\,d$
5	What is the equation for pressure	pressure = $\frac{\text{force normal to a surface}}{\text{area of that surface}}$	$p = \frac{F}{A}$
6	What is the equation for distance?	distance travelled = speed × time	$s = v\,t$
7	What is the equation of acceleration?	acceleration = $\frac{\text{change in velocity}}{\text{time taken}}$	$a = \frac{\Delta v}{t}$
8	What is the equation for resultant force?	resultant force = mass × acceleration	$F = m\,a$
9	What is the equation for momentum?	momentum = mass × velocity	$p = m\,v$
10	What is the equation for kinetic energy?	kinetic energy = 0.5 × mass × (speed)^2	$E_k = \frac{1}{2}m\,v^2$
11	What is the equation for gravitational potential energy?	gravitational potential energy = mass × gravitational field strength (g) × height	$E_p = m\,g\,h$
12	What equation links power, energy transferred, and time?	power = $\frac{\text{energy transferred}}{\text{time}}$	$P = \frac{E}{t}$
13	What equation links power, work, and time?	power = $\frac{\text{work done}}{\text{time}}$	$P = \frac{W}{t}$
14	What are the equations for efficiency?	efficiency = $\frac{\text{useful output energy transfer}}{\text{total input energy transfer}} = \frac{\text{useful power output}}{\text{total power input}}$	
15	What is the equation for wave speed?	wave speed = frequency × wavelength	$v = f\,\lambda$
16	What is the equation for charge flow?	charge flow = current × time	$Q = I\,t$
17	What is the equation for potential difference?	potential difference = current × resistance	$V = I\,R$
18	What equation links power, potential difference, and current?	power = potential difference × current	$P = V\,I$
19	What equation links power, current, and resistance?	power = (current)^2 × resistance	$P = I^2\,R$
20	What equation links energy transferred, power, and time	energy transferred = power × time	$E = P\,t$
21	What equation links energy transferred, charge flow, and potential difference?	energy transferred = charge flow × potential difference	$E = Q\,V$
22	What is the equation for density?	density = $\frac{\text{mass}}{\text{volume}}$	$\rho = \frac{m}{V}$

Put paper here

You will be provided with a *Physics Equations Sheet* that contains the following equations. You should be able to select and apply the correct equation to answer the question.

	Word equation	Symbol equation
1	pressure due to a column of liquid = height of column × density of liquid × gravitational field strength (g)	$p = h \rho g$
2	(final velocity)² – (initial velocity)² = 2 × acceleration × distance	$v^2 - u^2 = 2as$
3	force = $\frac{\text{change in momentum}}{\text{time taken}}$	$F = \frac{m \Delta v}{\Delta t}$
4	elastic potential energy = 0.5 × spring constant × (extension)²	$E_e = \frac{1}{2} k e^2$
5	change in thermal energy = mass × specific heat capacity × temperature change	$\Delta E = m c \Delta \theta$
6	period = $\frac{1}{\text{frequency}}$	
7	magnification = $\frac{\text{image height}}{\text{object height}}$	
8	force on a conductor (at right angles to a magnetic field) carrying a current = magnetic flux density × current × length	$F = B I l$
9	thermal energy for a change of state = mass × specific latent heat	$E = m L$
10	$\frac{\text{potential difference across primary coil}}{\text{potential difference across secondary coil}} = \frac{\text{number of turns in primary coil}}{\text{number of turns in secondary coil}}$	$\frac{V_p}{V_s} = \frac{n_p}{n_s}$
11	potential difference across primary coil × current in primary coil = potential difference across secondary coil × current in secondary coil	$V_p I_p = V_s I_s$
12	For gases: pressure × volume = constant	$p V$ = constant

Great Clarendon Street, Oxford, OX2 6DP, United Kingdom

Oxford University Press is a department of the University of Oxford.

It furthers the University's objective of excellence in research, scholarship, and education by publishing worldwide. Oxford is a registered trade mark of Oxford University Press in the UK and in

certain other countries

First published in 2019

British Library Cataloguing in Publication Data

Data available

978-1-38-200488-6

10 9 8 7 6 5 4

Paper used in the production of this book is a natural, recyclable product made from wood grown in sustainable forests.

The manufacturing process conforms to the environmental regulations of the country of origin.

Printed in Great Britain by Bell and Bain Ltd, Glasgow

Acknowledgements

The publisher would like to thank the following for permissions to use copyright material:

Helen Reynolds would like to thank her editors for their support and feedback, and her wonderful friends Michele, Rob, Lesa, Galla, and Bill for all their support, long walks and coffee. She would also like to thank Oleksiy, her dance instructor, for providing continuing encouragement for her writing and for her foxtrot.

Cover illustration: Andrew Groves

p9: Shutterstock.

Artwork by Q2A Media Services Inc. and OUP.

Although we have made every effort to trace and contact all copyright holders before publication this has not been possible in all cases. If notified, the publisher will rectify any errors or omissions at the earliest opportunity.